Swatch Reference Guide for Interior Design Fabrics

DEBORAH E. YOUNG

THE FASHION INSTITUTE OF DESIGN AND MERCHANDISING

FAIRCHILD BOOKS AN IMPRINT OF BLOOMSBURY PUBLISHING INC.

BLOOMSBURY

NEW YORK • LONDON • OXFORD • NEW DELHI • SYDNEY

Fairchild Books
An imprint of Bloomsbury Publishing Inc

1385 Broadway
New York
NY 10018
USA

50 Bedford Square
London
WC1B 3DP
UK

www.bloomsbury.com

FAIRCHILD BOOKS, BLOOMSBURY and the Diana logo are trademarks of Bloomsbury Publishing Plc

First published in 2017

Library of Congress Cataloging-in-Publication Data
Names: Young, Deborah E., author.
Title: Swatch reference guide for interior design fabrics / Deborah E. Young.
Description: New York : Fairchild Books, 2016.
Identifiers: LCCN 2016023957 | ISBN 9781501306006 (hardback)
Subjects: LCSH: Textile fabrics in interior decoration—Handbooks, manuals, etc. | Textile fabrics—Handbooks, manuals, etc. | Textile fibers—Handbooks, manuals, etc. | Swatches (Dry-goods)—Handbooks, manuals, etc. | BISAC: ARCHITECTURE / Interior Design / General. | DESIGN / Textile & Costume.
Classification: LCC NK2115.5.F3 Y68 2016 | DDC 746—dc23
LC record available at https://lccn.loc.gov/2016023957

ISBN: 978-1-5013-0600-6

Cover design: Eleanor Rose

Typeset by Lachina
Printed and bound in the United States of America

Contents

	Preface	xv
	How to Use the Swatch Reference Guide	xvii
	Acknowledgments	xix
Chapter 1	The Textile Cycle: Fiber, Yarn, & Fabric	1
Chapter 2	Fiber Classification: Natural Fibers	13
Chapter 3	Fiber Classification: Manufactured Fibers	25
Chapter 4	Fiber Classification: Synthetic Fibers	33
Chapter 5	Yarn Classification	45
Chapter 6	Plain Weaves	53
Chapter 7	Plain-weave Variations: Basket and Rib Weaves	63
Chapter 8	Twill Weaves	69
Chapter 9	Satin Weaves	75
Chapter 10	Complex Weaves	81
Chapter 11	Pile Weaves	87
Chapter 12	Weft Knits	91
Chapter 13	Warp Knits	101
Chapter 14	Minor Fabrications	105
Chapter 15	Dyed and Printed Fabrics	115
Chapter 16	Fabrics Defined by Finishes	123
Quick Reference Guide Tables		129
Swatch Boards		137

Extended Contents

Preface xv
- The Paradox of Sustainability xv
- The Objectives of the Text xvi
- The Study of Textiles xvi
- The Organization of the Text xvi

How to Use the Swatch Reference Guide xvii
- The Objectives of the Fabrics in This Reference xvii
- Building the Swatch Reference Guide xvii
  - *Instructions* xvii
  - *Swatchbook Assembly* xviii
  - *Pick Glass* xviii
  - *Swatch Boards* xviii
  - *Quick Reference Guides* xviii

Acknowledgments xix

Chapter 1 The Textile Cycle: Fiber, Yarn, & Fabric 1
- The Textile Cycle 1
- The Right Stuff 1
- The Interior Design Textile Industry 2
  - *Responsible Selection and Practices* 2
- Professional Textile Organizations 3
- New Textiles—New Expectations 4
- Basic Definitions 6
  - *The Physical Textile Cycle* 6
- Table 1.1 Basic Textile Definitions 6
- **Swatches 1–4** 6
- The Language of Textiles 6
- Table 1.2 Classification of Fibers 7
  - *Textile Performance Concepts and Properties* 7
- Activity 1.1 Research Project: New Textiles 11

Chapter 2 Fiber Classification: Natural Fibers 13
- Overview: Natural and Manufactured Fibers 13
- Cellulose Fibers 13
  - *Seed Fibers* 13
- **Swatches 4–6** 14
  - *Bast Fibers* 14
- **Swatch 7** 14
- **Swatches 8–9** 15
  - *Leaf Fibers* 15
- **Swatch 10** 15
  - *Other Cellulose Fibers* 15
- Protein Fibers: Wool and Silk 15
- **Swatches 11–16** 15
  - *Wool* 15
- **Swatches 11–12** 15
  - *Silk* 15
- **Swatches 13–16** 16
- Table 2.1 Properties Common to All Cellulose Fibers: Cotton, Linen, Ramie, Hemp 18
- Table 2.2 Cellulose Properties Study Guide 19
- Table 2.3 Minor Hair Fibers 20
- Table 2.4 Properties Common to All Protein Fibers 20
- Table 2.5 Properties of Individual Protein Fibers 21
- Table 2.6 Comparison of Protein Fiber Properties 21
- Activity 2.1 Swatch Page: The Diversity of Cotton 23

Chapter 3 Fiber Classification: Manufactured Fibers 25
- Manufactured Cellulose 25
  - *Rayon* 25
- **Swatch 17** 25
- Table 3.1 Properties of Individual Manufactured Fibers 26
  - *Bamboo* 27
- **Swatch 18** 27
  - *Modal* 27
- **Swatch 19** 27
  - *Lyocell* 27
- **Swatch 69** 27
  - *Acetate* 27
- **Swatch 20** 28
  - *Cocona®* 28
- Manufactured Protein 28

Swatch Reference Guide for Interior Design Fabrics

Swatches 18–24 28
Soy 28
Seaweed 28
Manufactured Mineral 28
Glass 28
Swatch 22 28
Metallics 28
Swatch 23 29
Stainless Steel 29
Mineralé 29
Activity 3.1 Swatch Page: The Diversity of Polyester 31

Chapter 4 Fiber Classification: Synthetic Fibers 33
Introduction to Synthetic Fibers 33
Nylon 33
Swatch 24 34
Acrylic 34
Swatch 25 34
Modacrylic 34
Polyester 34
Swatch 26 34
Swatch 19 34
Olefin 35
Swatch 27 35
Aramid 35
Swatch 28 35
Spandex 35
Rubber 35
The Burn Test 35
Table 4.1 General Properties of Synthetic Fibers 36
Table 4.2 Properties Specific to Each Synthetic Fiber 36
Table 4.3 Significance of Properties Common to All Synthetic Fibers 37
Table 4.4 Burn Categories of Fibers 38
Table 4.5 Burn Characteristics of Fibers 39
Activity 4.1 Fiber Burn Test 41
Activity 4.2 Fiber Application Assignment 43

Chapter 5 Yarn Classification 45
Yarn Classification 45
Filament Yarns 45
Spun Yarns 46
Table 5.1 Properties of Spun and Filament Yarns 46
Yarn Types 46
Novelty or Complex Yarns 46
Swatches 32–35 46
Yarn Twist 47
Table 5.2 Properties Affected by Analysis of Twist 48
Yarn Performance 48
Yarn Sizing 48
Yarn Count System 48
The Denier System 49
Microdenier/Microfiber 49
The Tex System 49
Swatches 29–35 49
New Innovations in Yarns 49
Sewing Thread 49
Table 5.3 Yarn Study Guide 50
Activity 5.1 Swatch Evaluation: Yarns 51

Chapter 6 Plain Weaves 53
Fiber and Fabric 53
Identifying Fabrics 54
Criteria for Fabric Identification 54
Table 6.1 Basic Weight Categories 55
Anatomy of a Woven Fabric 55
Qualitative Analysis of a Fabric 55
Plain Weaves 56
Swatches 36–45 56
Interior Design Focus: Plain Weaves 58
Window Treatments 58
Sheets/Bedding 58
Upholstery 58
Activity 6.1 Fabric Evaluation by Weight 59

Chapter 7 Plain-weave Variations: Basket and Rib Weaves 63
Basket Weaves 63
Swatches 46–47 64
Rib Weaves 64
Swatches 49–55 64
Table 7.1 Performance Expectations of Basket and Rib Weaves 64
Activity 7.1 Woven Fabric Analysis 67

Chapter 8 Twill Weaves 69
Twill Categories 70
Uneven Twills: 2:1 (1:2) or 3:1 (1:3) 70
Swatches 56–58 70
Even-sided Twills 70
Performance Expectations of Twill Weaves 71
Swatches 59–62 71
Activity 8.1 Twill Application 73

Chapter 9 Satin Weaves 75
Performance Expectations of Satin Weaves 76
Study Guide 76
Swatches 63–67 76
Table 9.1 Comparison of Basic Weaves 78
Activity 9.1 Weave Comparison Graph 79

Chapter 10 Complex Weaves 81
Slack-tension Weaves 81
Swatch 68 81
Dobby Weaves 81

Swatches 69–75 82
An Overview of Crepes 82
Crepe Fabrics 82
Extra-yarn Weaves
(aka Supplemental Warp or Weft) 82
Swatches 76–78 82
Jacquard Weaves 82
Characteristics of Jacquard Fabrics 82
Swatches 79–85 83
Double Weaves 83
Swatches 86–88 83
Leno Weaves 83
Swatches 89 83
Interior Design Focus: Complex Weaves 83
Window Treatments 83
Sheets/Bedding 83
Upholstery 83
Study Guide 83
Activity 10.1 Research Project: Storybook 85

Chapter 11 Pile Weaves 87
Construction of Pile Weaves 87
Performance Expectations of Pile Weaves 88
Swatches 91–97 88
Study Guide 88
Activity 11.1 Fabric Structures 89

Chapter 12 Weft Knits 91
Table 12.1 Comparison of Weaves and Knits 92
Fabric Gauge 92
Evaluating Stretch 92
Knit Quality Criteria 93
The Four Basic Knit Stitches 93
Classifications of Knit Fabrics 94
Basic Weft-knit Fabrics 94
Swatch 98 94
Table 12.2 Comparison of Weft
and Warp Knits 95
Rib Knits 95
Double Knits 95
Swatch 99 95
Pile Knits 96
Swatchs 100–102 96
Study Guide 96
Activity 12.1 Knit Fabric Analysis 97
Activity 12.2 Fabrication Worksheet 99

Chapter 13 Warp Knits 101
Tricot 101
Swatches 103–106 101
Raschel 101
Table 13.1 Contrast of Warp and Weft Knits 102
Table 13.2 Comparison of Tricot
and Raschel Knits 102
Swatches 107–110 102
Study Guide 102
Activity 13.1 Fabric Structures: Knits 103

Chapter 14 Minor Fabrications 105
Fabrics Made without Yarn 105
Swatches 111–113 106
Fabric Combinations 106
Swatch 105 106
Fabrics Made with Yarn 106
Swatches 116–118 106
Table 14.1 Lace Putups 107
Fabrics Made without Yarn or Fiber 108
Skins 108
Swatches 111–119 109
Interior Design Focus: Minor Fabrications 109
Window Treatments 109
Sheets/Bedding 109
Table 14.2 Minor Fabrications 109
Upholstery 110
A Brief Carpeting Primer 110
Table 14.3 Carpeting Fibers 110
Table 14.4 Carpet Structures 111
Activity 14.1 Carpet Analysis 113

Chapter 15 Dyed and Printed Fabrics 115
The Basic Dye Process 115
Stages of Dyeing 116
Fiber or Stock Dyeing 116
Swatch 120 116
Solution or Dope Dyeing 116
Yarn Dyeing 116
Swatch 121 116
Piece Dyeing 116
Swatches 120–122 117
Basic Dye Chemistry 117
Table 15.1 Properties of Dyes
and Pigments 117
Special Dye Processes 118
Color Management 118
Printed Fabrics 119
Swatches 120–137 120
Activity 15.1 Coloration Methods 121

Chapter 16 Fabrics Defined by Finishes 123
General Finishes 123
Aesthetic Finishes 123
Select Visible Aesthetic Finishes 123
Functional Finishes 124
Emerizing/Sueding 124
Swatches 138–145 124
Table 16.1 Fabric Selection
by Application 125
Activity 16.1 Client Project 127

Quick Reference Guide Tables 129
Table 9.2 Quick Reference Guide for the Three Basic Weaves 129
Table 10.1 Quick Reference Guide for Complex and Pile Weaves 131
Table 12.3 Quick Reference Guide for Weft Knits 133
Table 13.3 Quick Reference Guide for Warp Knits 135

Swatch Boards 137

Swatch Board Contents

Chapter 1 The Textile Cycle: Fiber, Yarn, & Fabric

Raw Fiber

1. Cotton

Yarn

2. Spun Yarn
3. Filament Yarn

Fabric

4. Muslin

Chapter 2 Fiber Classification: Natural Fibers

Cellulose Fibers

5. Cotton
6. Organically Colored Cotton
7. Flax
8. Hemp
9. Jute
10. Raffia

Protein Fibers

11. Wool
12. Mohair
13. Silk
14. Wild Silk
15. Silk Noil
16. Dupioni Silk

Chapter 3 Fiber Classification: Manufactured Fibers

Manufactured Cellulose Fiber

17. Rayon
18. Bamboo
19. Micromodal®
20. Acetate
21. Ingeo® Corn

Manufactured Minerals

22. Glass
23. Metallic

Chapter 4 Fiber Classification: Synthetic Fibers

24. Nylon
25. Acrylic
26. Polyester
27. Olefin
28. Nomex® Aramid

Chapter 5 Yarn Classification

Filament and Spun Yarns

29. Single Multifilament
30. Single Spun
31. Two-ply Spun

Novelty Yarns

32. Bouclé
33. Chenille
34. Chenille Damask
35. Tweed

Chapter 6 Plain Weaves

36. Voile
37. Filament Voile
38. Crushed Filament Voile
39. Gauze
40. Organdy
41. Organza
42. Gingham
43. Chambray
44. Shantung
45. Crash/Butcher Linen

Chapter 7 Plain-weave Variations: Basket and Rib Weaves

Basket Weaves

46. Monk's Cloth
47. Canvas

Rib Weaves

48. Iridescent Taffeta
49. Crepe de Chine
50. Broadcloth
51. Repp
52. Bengaline
53. Ottoman

Vertical Rib Weave

54. Pincord
55. Dimity

Chapter 8 Twill Weaves

56. Denim
57. Drill
58. Ticking
59. Herringbone
60. Houndstooth
61. Glen Plaid
62. Twill Plaid

Chapter 9 Satin Weaves

63. Satin
64. Crushed Satin
65. Antique Satin
66. Flannel-back Satin
67. Sateen

Chapter 10 Complex Weaves

Slack-tension Weaves

68. Seersucker

Dobby Weaves

69. Dobby—Filament
70. Dobby—Spun
71. Bird's-eye Piqué
72. Waffle Cloth
73. Goose-eye Twill
74. Satin Stripe
75. Crepe/Momie Weave

Extra-yarn Weaves

76. Extra-yarn Weave
77. Clip Spot
78. Dotted Swiss

Jacquard Weaves

79. Tapestry
80. Cotton Damask
81. Printed Damask
82. Flamestitch
83. Brocade
84. Brocatelle
85. Liseré

Double Weaves

86. Double Weave
87. Pocket Cloth
88. Cloqué
89. Matelassé

Leno Weaves

90. Marquisette

Chapter 11 Pile Weaves

91. Terry Cloth
92. Widewale Corduroy
93. Velveteen
94. Velvet
95. Panné Velvet
96. Voided Velvet
97. Grospoint

Chapter 12 Weft Knits

98. Jersey

Double-weft Knits

99. Double-knit Mattress Pad

Pile Knits

100. Sliver Knit
101. Fleece
102. Bonded & Embossed Pile Knit

Chapter 13 Warp Knits

103. Tricot
104. Brushed Tricot
105. Sueded Tricot
106. Minky
107. Tulle
108. Raschel Lace
109. Raschel Crochet
110. Point d'Esprit

Chapter 14 Minor Fabrications

Nonwovens

111. Imitation Suede
112. Needlepunched Felt

Film

113. Pleather
114. Oilcloth

Fabric Combinations

115. Quilted

Embroideries

116. Embroidered Eyelet

Tufted

117. Tufted Chenille

Lace

118. Venise Lace

Skin
119. Leather

Chapter 15 Dyed and Printed Fabrics

Stages of Dyeing
120. Fiber/Stock Dyed
121. Yarn Dyed
122. Piece Dyed
Classic Print Designs
123. Calico/Ditsy
124. Paisley
125. Toile de Jouy
Non-Classic Designs
126. Direct Print
127. Blotch Print
128. Overprint
Better Quality Prints
129. Discharge Print
130. Flock Print
131. Flocked Dotted Swiss
132. Velvet Burnout
133. Flat Burnout
134. Duplex Print
Resist Prints
135. Tie-dye
136. Batik
137. Ikat

Chapter 16 Fabrics Defined by Finishes

Napping
138. Flannel
139. Flannelette
Calendering
140. Glazed Chintz
141. Moiré Taffeta
Embossing
142. Embossed
Pleating
143. Pleated
Emerizing/Sueding
144. Suede cloth
145. Moleskin

Preface

We have always had an intimate relationship with textiles, holding them next to our skin, allowing them to comfort, protect, and reveal us. They are an integral part of our everyday lives. We sleep under them, sit on them, and walk on them. Rumpelstiltskin believed that the spinning of yarn was his path to riches. Harry Potter used a cloak to hide in plain sight, while Aladdin allowed his carpet to carry him away. The three fates even determined our life spans and destinies by measuring and cutting the threads of our lives.

Our common language is rich with textile metaphors, as we discuss the fabric of our lives, following the thread of a story, or spinning a yarn. We might even put a spin on an idea or event, reframing the actual facts. When we speak of someone's emotional distress, we might suggest that they are high strung, fraying at the edges, tightly wound, enmeshed, unraveling, or coming apart at the seams. If we find someone that we have a kinship with, we could find that we are cut from the same cloth. In the romantic realm, we have sometimes been known to weave a magic spell over the object of our desire. Smooth as silk is a term used to express something that was easy to accomplish. And following the thread of an online conversation is part of our everyday lives. Textiles have ever been a symbol of our cultures and humanity. And yet the science behind the textiles we take for granted continues to elude us.

The intention of this book is to demystify the science and make it useful for the interior designer in a meaningful way: to provide them with the essentials to make appropriate choices.

The text uses simple, direct language that is not specific to textile scientists but rather language that is familiar to the industry at large. The interior design trade requires textile science to achieve the appropriate performance of the product. The goal of this book is not to soften, but to focus the science in a way that is more accessible to the designer. Instead of an in-depth analysis of the molecular structure of a fiber, the text focuses on the relevant performance expectations for the best possible placement of products in an environment.

A solid understanding of basic textile science will assist professionals in making better choices in fibers and fabrics for their chosen end products. This text strikes the necessary balance between scientist and designer. It culls the information available to the textile scientist and presents only the material directly relevant to the designer.

This book brings all the elements of a textile together into a common place. With all the information in one location (swatches and text), students can spend less time attempting to connect the dots and more time applying the concepts.

The swatches in this text are shopped from the current marketplace, including jobbers, wholesalers, and retailers. As such, they are a reflection of today's market and right on trend.

The Paradox of Sustainability

Sustainability has become a hot topic. It has also become a very confusing ethical dilemma and even more so in the world of interiors. Does recycling mean sustainable? Is the fiber sustainable if it grows without pesticides? Or is it enough to simply reduce water usage? And is the fiber still sustainable if it is grown without pesticides but dyed in a chemical bath, finished with formaldehyde, and then shipped overseas in plastic? While each of these gestures is a giant leap toward more socially responsible behavior, these questions represent only a small number of complex issues inherent in the current state of defining sustainable methods and practices today.

To properly incorporate earth-friendly practices, a study of economic, environmental, and social practices must be examined in all aspects of a product's life cycle. It is not enough to just make sure that pesticides are not used; the life cycle of the product must be tracked from creation through its ultimate disposal.

The sustainability paradox in the fashion world (including the fashion of interiors) is that sustainability is about longevity, but fashion is about perpetual newness and change. These industries (textiles and design) do have a genuine desire to incorporate green behaviors, but true sustainability does not end at purchase. In order to completely close the loop, consumers must also commit to reusing, recycling, repurposing, and reducing consumption.

In an attempt to address green textiles in a meaningful way, qualifying swatches have been classified into four distinct areas. Icons have been added to the swatch boards that represent the following sustainability criteria:

1. Biodegradable fabrics
2. Potentially recyclable **or** recycled materials
3. Sustainable fabrics
4. Fabrics made within a closed loop system

B Biodegradable is usually a term used for natural fabrics that will reasonably quickly break down by bacteria, fungi, or other simple organisms. (Even polyester will eventually biodegrade. The difference is the time required to break down, and polyester is generally not considered biodegradable because of the centuries that would be required to break down). Biodegradable fabrics will be notated on the swatch boards with this icon. It should be noted that biodegradability does not occur in a landfill.

Recycled materials and materials that can be recycled have been notated on the swatch boards with this icon. Note that although many of these fabrics have not yet been recycled, any 100 percent polyester or nylon fabric can be shredded down to make other fleecy type fabrics. Natural fibers are also making progress in recycling. It is the responsibility of the consumer to make sure these fabrics do not end up in landfills but are properly donated to collection systems, such as Patagonia's Common Threads for polyester or Cotton Incorporated's Blue Jeans Go Green. Through the latter, donated jeans find a new life as insulation.

Despite their popularity, fiber blends are problematic to recycle. The current addition of elastomeric fibers, such as spandex, makes recycling fiber blends an even more complex and difficult issue. Through a process of disaggregation, new life is being found for some of these blends.

Sustainable fabrics are made from materials that can be replanted or recreated without depleting natural resources.

Closed Loop system means that the solvents used in the process of making these fibers is completely (or almost completely) reclaimed, without damage to our environment.

The Objectives of the Text

- Create awareness of the diversity of textiles available.
- Provide a basic working knowledge of textile composition, function, and application. This will enable the student to have the information necessary to make informed decisions regarding textiles and to communicate with industry professionals.
- Demonstrate correct use of textile terminology.
- Differentiate between two critical concepts:
  - The difference between fiber and fabric.
  - The difference between fabric structures, most commonly weaves and knits.
- Explain production processes and how they impact the fabric. This would include potential product performance and selection, based on fiber, yarn, fabrication, coloration, and finishes.
- Differentiate fiber classifications, yarn types, and fabrication methods, and determine how different fabrics will perform for a specific end use.
- Demonstrate the selection of appropriate components of a product with respect to compatibility with one another and with the desired result.

The Study of Textiles

Ultimately the study of textiles will help designers to make informed decisions. For example, silk is a lovely and desirable fiber with a pleasant hand and excellent drape. With this in mind, one might come to the conclusion that it would be a good choice for a drape. As it is damaged by the sun, it is usually a poor choice unless it is properly lined and protected from the sun's harmful rays. The fiber content of linings, interfacings, backings, and trims must also be factored in. For the sake of compatibility, do they have to be made of the same fiber? Often polyester is used as a lining for a drape because of its superior sunlight resistance. Silk and polyester are generally not considered compatible, and yet they can live together. These are complicated questions that the designer should be able to answer before placing a silk drape in a sunny window.

The problem lies in the disparity in care between polyester and silk. They shrink at different rates, and they have different ironing requirements. Cotton and polyester are used together even more often, and yet these fibers are even more remarkably different. Fiber mixing must be done judiciously. This example represents a tiny fraction of the myriad decisions that designers will face in their career. A diligent study of textiles will provide the knowledge and confidence to make more informed and reliable choices.

Cost is also a huge factor. Interior fabrics are much more expensive than fashion fabrics. Mistakes in fabric choices can dramatically impact the financial bottom line. In fact, one of the few variables in the cost of a product is the textile itself. The study of textiles will assist in learning to shop for fabrics appropriate to a given use, design, or product silhouette. Knowledge of textiles will also assist in finding specific textiles for less money or finding ways to reduce the bottom line. In addition, the information gained will assist in quality-control recognition and component compatibility.

The Organization of the Text

This text is organized to follow the natural and logical sequence of events that occur in the production of a textile. The first four chapters deal with fibers: natural fibers (cellulose and protein) followed by manufactured and synthetics. The next chapter addresses the relevance of yarn constructions. The body of the text is devoted to the identification and articulation of fabrics by structure and name. The final chapters address the dyeing, printing, and finishing of fabrics. Each chapter is punctuated with representative examples of swatches to reinforce the subject of the chapter.

Fibers: Natural & Man-made

↓

Yarn Constructions: Spun & Filament

↓

Fabrics: Woven, Knit, & Minor Fabrications

↓

Dyes & Prints

↓

Finishes

↓

End Products

How to Use the Swatch Reference Guide

The Objectives of the Fabrics in This Reference

Fabrics in this text are logically organized into groups by their similarities. The first chapters are the fiber chapters, and swatches are included for their fiber content. The swatches in Chapter 5 are presented to illustrate yarn types. In each subsequent chapter, the fabrics are presented as relevant samples of classic fabrics made of the structure at issue.

Fabrics have been around for a long time. Because manufactured fibers had not yet been created, most fabrics were originally invented with one of the big-four fibers: wool, flax, silk, and cotton. This text will often refer to fabrics as cotton-like or from the cotton fiber family, which relates to their first incarnations and to their hand. Today, a cotton-like fabric might be made entirely of polyester. (This would be an example of polyester imitating cotton, as it often does.)

As a final note, fiber and fabrics are not inextricably linked. Gingham is a fabric that can be made of cotton, polyester, rayon, or even a blend. Think of fiber and fabric as first and last names, such as Amanda Connell. Not all Amandas have the last name of Connell; there could be an Amanda Andrews or even an Amanda Smith, and there are certainly other Connells. Similarly, not all cottons are denim: Cotton could be a jersey, velveteen, corduroy, or many other fabrics. And today, not all denims are cotton either. Think of fiber and fabric as independent variables.

Building the Swatch Reference Guide

Your first task will be to build the Swatch Reference Guide. One of the things that you will notice is that text is provided for each swatch; you need only mount the swatches. Although the book is organized into logical and sequential chapters, the instructor may well use swatches out of order. It is advised that you construct the entire Swatch Reference Guide during the first week of class, mounting all of the swatches at once. This will enable you to have a complete resource of swatches at your fingertips for the instructor to draw upon to illustrate ideas.

Instructions

Begin by identifying the materials in this package:

- 1 bag of fibers and two yarns: Swatches 1–3
- Four Bundles of Swatches: A, B, C, & D
    - A has Swatches 4–35
    - B has Swatches 36–85
    - C has Swatches 86–105
    - D has Swatches 106–145
- Swatch boards to mount the swatches
- Pick glass
- The text and swatch boards are shrink-wrapped together. Insert the swatch boards at the end of each chapter. There are five blank swatch boards that will allow you to expand this fabric reference with your own selection of fabrics.

Based on the availability of some fabrics, there are a few minor color variations in the swatches presented. In all cases, the swatches have the same character: They share the same fiber content, yarn construction, count, weight, stage of dyeing, and finishes.

All of the swatches are numbered and correspond to references in the text. The swatches are bundled in the order that they appear in the text. Do not remove the rubber bands on the bundles until you are ready to assemble the Guide. Take swatches from the top of the bundle and keep the stack face up.

Note that some of these swatches are lightweight and stick together, due to static cling. Check the front and back of each swatch to prevent this.

You will also need the following (not included):

- Pick needle. It is often helpful to have a pick needle to assist in your analysis of fabrics, particularly when it comes to counts or pick outs (analyzing the structure of the fabric).
- Roll of double-sided tape.

Swatchbook Assembly

Open the baggie first and mount the fiber to the box for Swatch 1 with double-sided tape. Place tape in the middle of the box, and cover the tape with fiber.

Mount the blue cotton yarn to the box for Swatch 2.

Mount the pink filament yarn to the box for Swatch 3.

Now begin mounting the swatches in the four bundles sequentially in the book. Open swatch bundle A and discard both the card stock on top and the marker paper directly below it. Your first swatch is muslin, Swatch 4. Place one inch of the double-sided tape horizontally across the top of the swatch box, and then place the swatch on top of the tape. This way, you can flip up the swatch to observe and feel both the front and back.

As you mount the swatches to the swatch boards, verify that the swatch matches the description included on the board. Rely on fabrics that you already know, such as denim or velvet. Check that you are on the right number as you get to the end of each bundle. If two swatches appear the same, you have been gifted with a duplicate. Do not add it to the book. Take this one out and put it aside.

Pick Glass

Verify that you have a pick glass in your kit. Open the pouch and unfold the glass completely. Notice that your pick glass is a one-inch square. Thread counts are calculated in this one-inch square.

Look through the glass to the ruler below, and look at the fabric on your sleeve to get used to the pick glass. You will use this instrument throughout your study of textiles, so have it with you for every class.

Swatch Boards

While most facts are provided for each swatch in the text, there are instances where the information for a specific swatch field has been intentionally left blank. This corresponds directly to activities in the text or in-class activities provided by your instructor. You will be required to determine and fill in the results.

Additionally, the facts provided in the yarn construction category are simplified. Unless otherwise stated, it is safe to assume that the yarn type is single, unless identified as plied. In the case of filament yarns, one can assume multifilament, unless otherwise noted, since this is most often the case.

Some criteria are present only when it is particularly relevant. Finishes, for example, is a missing criteria for most fabrics when it is not an aesthetic or visible finish. This does not mean that there is no finish on the fabric. This text recognizes that most fabrics have a dozen finishes on them before they are seen by the consumer; it means that there is no visible or discernible finish.

One of the possibilities under the coloration criteria is bleach. As dye is transparent, white is not achievable without a bleaching or color removing agent. In this text, bleach is used as a generic term to represent color removers. The text acknowledges that chlorine bleach is usually rejected as it is caustic and destructive to the fiber.

Quick Reference Guides

You will find Quick Reference Guide tables for Basic Weaves (Table 9.2), Complex and Pile Weaves (Table 10.1), Weft Knits (Table 12.3), and Warp Knits (Table 13.3) preceding the Swatch boards. Please remove each table and place it within its respective chapter for easy referencing.

Acknowledgments

An undertaking of this sort is truly a collaborative project, and there are many people I wish to thank for their patience, encouragement, and support. I am so grateful to Carol Shaw Sutton, who initially nurtured a love of textiles and helped me to see the world through fiber eyes. Barry Jo Sims was an essential part of my education for this subject. Amanda Starling provided the motivation for the project. Rubin Schubert, Mauricio Pena, Marco Ramirez, and the crew at Ragfinders were of enormous assistance by providing just the right choices for many of the exciting fabrics found in this resource. Thanks also to Gina Rios of Designer Fabric Warehouse for her assistance in this massive scavenger hunt for fabrics. My colleagues and friends, Ben Amendolara, Cassandra Durant-Hamm, and Judy Picetti, were always willing to contribute when I called on them.

Thanks to German Salgado at Santa Ana Cutting for his vigilance in ensuring that all swatches are present. I also wish to extend a heartfelt thanks to the Fairchild Books team for their initial vision and realization of the final product. This book would not have come together without the concerted efforts of Noah Schwartzberg, acquisitions editor; Amy Butler, development editor; Edie Weinberg, art development editor; and Claire Henry, production editor. I am eternally grateful for the patience of my family during the course of this project: Amanda, Mariah, Diana, Mike, Kim, and Alyssa. My four-legged family, Gwennie and Morgie, have been far and away the most patient with the process. Most of all, I want to thank Jim Young, my tireless husband, who continues to support me with this process, travel with me on every wild goose chase, and most of all, for his unending faith in me.

Thanks to the reviewers for their invaluable feedback: Lisa Curwen, Rocky Mountain College of Art and Design; Celeste Day-Drake, FIDM; Susan Eid, Montclair State University; Mark Gower, Regent's University London (UK); Darla Green, Johnson County Community College; Trudy Landgren, St. Catherine University; Shauna Morris, Webster State University; Tracey Seibert, University of Cincinnati; and Barry Jo Sims, FIDM.

The Publisher wishes to gratefully acknowledge and thank the editorial team involved in the publication of this book:

Acquisitions Editor: Noah Schwartzberg

Development Editor: Amy Butler

Assistant Editor: Kiley Kudrna

Art Development Editor: Edie Weinberg

In-House Designer: Eleanor Rose

Head of Production (Academic): Jo Murphy

Production Manager: Claire Henry

Project Manager: Morgan Ewald, Lachina

CHAPTER ONE

The Textile Cycle: Fiber, Yarn, & Fabric

Reference Swatches

1. Cotton | 2. Spun Yarn | 3. Filament Yarn | 4. Muslin

Textiles are ubiquitous, playing an integral part of our everyday lives. The paradox is that they are often invisible to us. We seldom notice the carpets that we walk on, the drapes of the rooms we pass through, or the upholstered seats of the cars that we drove to work. We can be oblivious to the bed linens that we crawled out of, the towels that we used in the morning, the chairs in the offices of businesses that we visit, or the clues we receive as to occupation or personality of the people we encounter in our day due to the garments or uniforms they wear. We take the textiles of our lives for granted, yet we would certainly notice their absence, especially in the environments in which we live. Without textiles for interior design, life would be louder, colder, and harder edged. It would be less friendly.

The Textile Cycle

The discovery and development of textiles—harvesting the fibers, spinning fibers into yarns, weaving the fabric, and sewing the fabric into the final product—was one of mankind's earliest technical achievements, right after taming fire and mastering stone tools. After 20-thousand-plus years of textile history, the basic processes for producing textiles have not changed. Fibers still need to be harvested and spun into thread or yarn. Yarns have to be manipulated on some type of loom structure to create fabric. And fabric is used to furnish our world.

To be sure, the Industrial Revolution and mechanization in the 1800s and the development of manufactured fibers in the last century brought new uniformity and speed to the production process. But it's our ingenuity and drive to produce stronger, cheaper, better, and more beautiful fabrics that make the field of contemporary textiles so exciting and diverse. With textile technology, the modification and creation of new fibers and fabrics seems limitless. For today's designer, it is no longer enough to select a textile simply for its hand, drape, or color. Today's consumer wants performance—fabrics that won't shrink, wrinkle, fade, or soil. And to ensure a sale, the fabric should also have some interesting *other* ability, like antistatic properties. In the current marketplace, we can actually meet these demands, and more.

The Right Stuff

A proper study of textiles for interior design culminates in understanding one essential concept: an understanding of and proper selection of the right fabric for a given product. This involves responding to questions, such as "Is this fabric strong enough for the sofa? Will this drape actually drape? Will this fabric keep its shape and appearance for a reasonable length of time?" Making appropriate fabric choices requires a thorough knowledge of the science of textiles. Understanding the hygroscopic, thermoplastic, electrical retention, or hydrophobic qualities of a fiber or fabric is essential for today's designers. This text is devoted to assisting the designer in making connections between textile science and interior design, supporting design with the practical application of relevant science.

The study of textiles begins with the smallest visible part of a textile, which is a fiber, and follows the actual textile cycle through to the final step, finishing. With increasing demand for more versatile and functional fabrics, chemical and mechanical finishing have become major areas of interest within the textile world, unlimited in their commercial potential and application. Often, knowledge of textiles begins and ends with a rudimentary idea of fiber content for both consumer and designer, but this is no longer enough information to make informed choices. Today's textiles are much more sophisticated, with so many areas of scientific possibilities. One segment of the textile industry of specific interest to interiors is devoted to the science of stain removal and stain resistance. This is particularly relevant in interiors, as laundering products is often not practical. It is difficult to throw the sofa into the washing machine.

Stain removal may not seem the most exciting segment of the industry, but when your pen has leaked ink onto your sister's new white sofa, it certainly becomes a compelling subject. (Hair spray will usually remove that ink and get you out of trouble.) This and other new developments in textile science are moving the textile industry into fascinating new realms.

The Interior Design Textile Industry

Fabrics for interiors continue to enjoy a large portion of the textile industry at large. Forty percent of the contemporary textile industry is devoted to fabrics for interior designers, another 40 percent supports the fashion industry, and the remaining 20 percent is allocated to industrial applications, such as filters, geotextiles, and the medical, space, and automotive industries.

There are five broad classifications of fabrics used for interior design:

- Upholstery fabrics, including pillows and cushions. This includes all elements used in the process of upholstery, such as fabrics, filling, linings, straps, backings, and trims. Compatibility of all of the elements must be considered.
- Window and wall coverings, including drapes, curtains, sheers, blinds, and shades. They can be made of anything, from a fine silk to a heavy jute, and often incorporate textured materials, such as metallics and grasses.
- Soft floor coverings, including everything from wall-to-wall carpeting to small scatter rugs. An understanding of padding and underlay is also essential.
- Household textiles, incorporating bedding, towels, table linens, napkins, and runners.
- Institutional textiles, including the same items as above but with more stringent durability requirements.

Responsible Selection and Practices

Sustainable practice involves creating good design that supports healthy economies and seeks to eliminate social inequities. In order for the industry to remain viable, it is incumbent on present and future designers to incorporate sustainable practices, upcycling, and recycling into their designs.

Sustainability is the ability to continue a defined behavior. It is how biological systems continue indefinitely.

Sustainability includes:

- sustainable biological systems, such as healthy wetlands and forests
- built-in continuity in production
- reduction in pollution
- business methods and practices that do not exploit people or natural resources
- responding to the lifestyle needs of today's consumer without jeopardizing the future
- longevity in design instead of disposal
- reduced energy and materials use
- consideration toward production, consumption, and ultimate disposal
- workers and animals treated with dignity and respect

Responsible design can include the use of **organic materials**. Organic fibers are created without pesticides or harm to the environment. Cotton is currently the only fiber that can be certified as organic.

True sustainability includes an analysis of the Life Cycle Assessment, **LCA**, which is a process for evaluating all of the environmental aspects of a product through all stages of its life cycle. Sometimes called **Cradle to Grave** analysis, this concept examines a product from its creation to disposal. **Cradle to Cradle** (C2C) uses environmentally intelligent design to examine the overall impact of production. Ultimately, cradle to cradle suggests that a product at the end of its life cycle should be recycled into a new product, with no waste, and no trash.

Greenwashing includes the use of deceptive practices in which a company, government, or other group promotes green-based or environmentally friendly business practices but actually operates in a way that is damaging to the environment or in an opposite manner to the goal of the announced initiatives.

Professional Textile Organizations

Numerous professional organizations assist in defining and setting standards for sustainability and best environmental practices in textile production.

AATCC—The American Association of Textile Chemists and Colorists is a technical and scientific society of members from more than sixty countries. The AATCC is a Standards Developing Organization (SDO) that has developed testing and evaluation procedures for the analysis of the physical properties of textiles. Since 1921 the AATCC has developed more than two hundred test methods and evaluation procedures.

ACT Standards—The Association for Contract Textiles makes fabric specification easier; they created standard guidelines for flammability and four aspects of fabric durability: wet and dry crocking, colorfastness to light, physical properties, and abrasion (Figure 1.1).

Figure 1.1 ACT icons identifying performance characteristics. From top—Flammability, Wet and Dry Crocking, Colorfastness to Light, Physical Properties, General Contract Upholstery, Heavy Duty Upholstery

ASTM—American Society for Testing Materials is an international standards organization that develops and publishes voluntary technical standards for a wide range of materials, products, systems, and services.

Blue Jeans Go Green™—Old denim gets upcycled into Cotton Incorporated's UltraTouch™ Denim Insulation made from 80 percent postconsumer recycled denim, providing a portion to communities in need, each year. This system keeps old denim out of landfills.

Bluesign®—Bluesign is a system for sustainable textile production and a third party auditor. Harmful substances are eliminated from the beginning of the manufacturing process. Control standards are set for environmentally friendly and safe production. This not only ensures that the final textile product meets very stringent consumer safety requirements worldwide but also provides confidence to the consumer to acquire a sustainable product. The Bluesign Pledge is to reduce, repair, reuse, recycle, and reimagine. All companies are audited and monitored.

Fair Trade—A social movement with two goals: (1) helping producers in developing countries achieve better trading conditions and (2) promoting sustainability.

FLA—Fair Labor Association creates lasting solutions to abusive labor practices by offering tools and resources to companies, delivering training to factory workers and management, conducting due diligence through independent assessments, and advocating for greater accountability and transparency from companies, manufacturers, factories, and others involved in global supply chains.

GOTS—Global Organic Textile Standards are a set of processing standards for organic textiles and fibers along the entire textile supply chain that ensure environmentally and socially responsible manufacturing. Producers must meet strict GOTS criteria for certification.

ISO—The International Organization for Standardization is a worldwide federation of national standards bodies. This organization provides certification to ensure companies doing business internationally adhere to standards and practices in production.

LEED—Leadership in Energy and Environmental Design is a third party verification and certification system to verify green design. The points earned in the LEED system determine the ultimate rating: Certified, Silver, Gold, or Platinum. The system assures resource efficiency: less water and energy usage and reduction of greenhouse gas emissions. Using post-consumer recycled fabrics qualifies for LEED credit.

NCIDQ—The National Council for Interior Design serves to identify to the public those interior designers who have met the minimum standards for professional practice by passing the NCIDQ Exam. The core purpose is to protect the health, life, safety, and welfare of the public by establishing standards of competence in the practice of interior design.

TFPIA—The Textile Fiber Products Identification Act stipulates the requirements for all textile labeling. This act determines that generic fibers must be listed first, followed by the trademark, and that fibers must be listed in descending order of content percentage (much like food labeling).

New Textiles—New Expectations

As visually stimulating and tactile as textiles are, they are even more exciting from a technological perspective. Consumers want high performance and low maintenance; they want textiles that can do tricks. The field is an exciting frontier. Space exploration, military, and medical research programs, along with the technology industries, have driven some of the most startling innovations in textile science. Often not directly inspired by, or even created for, interior design, these innovations trickle down to the world of couture and ultimately, into our homes. All it takes is a little creative thinking to make the leap from battlefield military to fashion couture and innovative interiors. Savvy designers use these new developments to meet the market demand for newer, innovative products.

Both military researchers and the medical field are rapidly producing advancements, like sensors that record and transmit information, such as heart rate, blood pressure, and insulin levels, to your doctor. While most of these innovations are currently being explored in the apparel industry, it does seem that they could be made even more relevant in interior applications such as bedding. In 2001, Fuji Spinning Company in Japan developed a T-shirt that provides your recommended daily allowance of vitamin C. Through a process called *microencapsulation*, your body slowly absorbs the vitamin transdermally (through the skin), just by wearing the shirt. The shirt continues to administer vitamins through as many as thirty to forty washes. Once again, transdermal applications of this sort might be even better suited to sheets or pillows.

Using the same technology, you could add many different medications to a textile. Consider a scarf that provides relief for headaches, gloves for arthritis sufferers, bedsheets with sleep inducers for insomniacs, or a textile for Alzheimer's patients. Using textures, sounds, and smells, a scarf has been created to calm autistic children. Each of these exist in today's technology, transforming the way health care will be delivered. Diagnosis, treatment, and follow-up can be conducted remotely. As fascinating as these possibilities are, they raise some provocative ethical questions as well. When these textiles become mainstream technology, will you need a prescription for your sheets or clothing? Will they have an expiration date? Will there be a black market for medicated sheets and underwear? These issues and many others will be facing the contemporary designer.

On the military front, invisibility in textiles is being explored, as well as GPS systems (global positioning systems), uniforms that will administer antibiotics when necessary, and textiles that stiffen to act as a splint as required, for combat injuries. Adapted to interiors, these textiles will be able to change their basic properties at will.

In another example of fiber-forward thinking, researchers are experimenting with spider silk (spider webs) because of its extreme strength (six times stronger than steel) and incredibly small size (eighty times thinner than a human hair). Spider silk is so strong that if you were to spin a strand of yarn to the diameter of a pencil, the tensile (breaking) strength is such that it could stop a 747, in flight. In manufacturing, fiber strength is critical because the stronger the fiber, the less that is needed for a particular use. Strength also relates to extended wear. Additionally, the spider silk is resistant to water, cold, heat, and fire.

Consider the use of spider silk for contract upholstery or other applications that have excessive use.

To date, the cultivation of spider silk has been problematic because the spiders will not cooperate. Unlike silkworms, spiders are territorial, and they recycle their proteins (that is, eat their webs and take them with them). Researchers have had to get creative in the cultivation of spider silk. Experiments are being successfully conducted in crossbreeding the DNA of spiders with potatoes, corn, and even goats. Herds of spider-goats produce milk that provides us with really strong fibers.

The following are some other high-performance textiles that are pushing the envelope of textile technology:

- Antibacterial textiles (no bacteria means no odor). Much experimentation is being successfully conducted in this arena. This could be exciting in the interior design world, where it is more difficult to clean products. Interior textiles are often dry clean or spot clean. This will lead to cleaner, odor-free, and more sanitary products.
- Textiles with living bacteria that will eat any spills or perspiration. The effect is a self-cleaning textile, which will be incredibly useful in upholstered furniture. As the bacteria are live, they must be fed (to keep them alive), so although you may not have to wash this textile, you might have to feed it on a regular basis!
- Super socks that are full of caffeine. They ramp up your metabolism to improve athletic performance. Imagine wearing your morning coffee—no need to stop and pick it up on the way! Extending this to interiors, this technology could be added to your towels and washcloths.
- Textiles that change color with temperature—or that change color and pattern with your mood (remember mood rings?). This is also being done with wallpaper (it changes pattern or color according to your whim). The wallpaper will become a microcomputer. Imagine a room that exhibits cool colors when the temperature is warmer and warm colors when the temperature is cooler.
- Textiles that change color with the presence of odorless pesticides or gases—great for detecting these dangers in your children's play areas.
- Window curtains that act as solar panels and power your house.
- Power Felt is a technology that can generate electricity from your own body heat. You can then throw your iPod or cell phone into your purse or bag and have it charged up for when you need it next.
- Textiles that adjust to your body temperature and regulate body humidity, cooling you when hot, warming you when cold. Using Thermocules™ (by Outlast Technologies), there are sheets that read your body temperature and self-regulate, creating adaptive comfort; you do not need to throw the covers on and off all night.
- Hoodies with cell phones or MP3 players built directly into a cuff or the hood. This technology has been extended to carpets and upholstered furniture. Programmed by touchpads under the carpet, adaptive rooms welcome you by adjusting the thermostat and turning on and off lights and music. Chairs will massage and warm you.
- A "smart bra" that turns into a sports bra by increasing its support as you begin to run, and then relaxing when you relax. Imagine extending this technology to protect mattresses or sofas when children are jumping on them.
- Smart textiles that transmit your personal vitals through the use of a smartphone. These are already in use in the medical profession, recording and transmitting sugar levels, heart rate, blood pressure, and even hormone levels. Ultimately, we will be wearing our computers and cell phones!
- Textiles that play movie trailers or short videos, perhaps turning your bed or walls into an electronic billboard.
- X-STATIC® (produced by Noble Biomaterials, Inc.) is incorporating silver fibers into nylon fabrics. The silver creates a static charge that assists in sanitizing garments without washing them. The effect is self-cleaning, and it is being used to make underwear, socks, shirts, hats, etc. Imagine—you can own just one pair of panties that you don't have to wash! Once again, this technology really does adapt well to interiors. Imagine a self-cleaning sofa, carpet, or even bedding.
- Scratch-and-sniff wallpaper that smells like bananas or another favored scent.
- Textiles with permanent scents or perfumes to enhance your experiences, perhaps used for the towels in a spa.
- Energear™ (by Schoeller) that converts your own energy back to you for improved regeneration, reduced fatigue, and quicker recovery time. Use this for your favorite chair to regenerate at the end of the day.
- Allergy medicines delivered remotely.
- Hug Pillows are two pillows connected remotely and shared with a couple separated by geography. When

one person hugs their pillow, the other pillow *glows,* remotely sending the hug to the other party.

- Microbial cellulose allows you to "grow" a chair or lamp.
- This is just the beginning, there are so many more ...

The technology is all very exciting, but first we need to appreciate some more fundamental concepts.

Basic Definitions

The first step in understanding textiles is mastering the vocabulary. Let us begin with some basic definitions that break down the language into simple terms so that you can begin speaking the language of textiles today (Table 1.1).

The Physical Textile Cycle

A **fiber** is the smallest visible part of a textile, a single hairlike strand. A fiber can be either staple or filament in length. **Staple fibers** are short—only inches long. All natural fibers are staple except for silk, which is nature's only filament. **Filament fibers** can be miles long and include both manufactured and silk fibers.

Cotton is staple, with short fibers, and so is wool. Acrylic is constructed as a man-made filament but is often cut to staple length, particularly when it is imitating wool. Likewise, in a blend such as polyester and cotton, the polyester would first be chopped into staple lengths for easier blending with the staple cotton fibers and for an all-over cotton-like hand. Although filament fibers can be cut to staple length, staple fibers cannot be made into long filaments.

The swatches in this chapter illustrate the different stages of a textile, from the raw fiber through the most common yarn types, spun and filament, and finally to the simplest fabric made from these fibers.

Swatch 1 is a staple cotton fiber. Swatch 2 is a spun yarn made of staple cotton fibers. Compare this with Swatch 3, which is filament polyester. Swatch 4 is a simple fabric made of staple fibers. These swatches represent the textile cycle: Fiber is harvested from the plant, spun into a yarn, and the yarn is used to construct the fabric. Note the simple difference in length of the fibers in the two yarns (Swatches 2 and 3). In Chapter 5, we will further explore these differences by comparing two fabrics identical in structure and fiber: one made of filament fibers and the other made of staple fibers.

[Reference Swatches 1–4]

The Language of Textiles

Before going any further, the core language of textiles needs to be introduced. This text will allow you to understand in detail the inherent performance properties of each fiber, yarn, and fabric construction. Familiarity with performance concepts and properties helps designers to determine the specific advantages and disadvantages a fabric will bring to its end use. In essence this knowledge reduces the possibility of making poor fabric choices for a particular product or end use.

Here is an example: We know that linen wrinkles horribly. By understanding the properties of and selecting a particular yarn or fabric construction, adding a select finish, or blending fibers, we can easily create a linen drape with reduced wrinkling.

The question is which fiber/fabric combination is best for a specific purpose? This will be answered by studying the chemical bases of fibers, (Table 1.2), along with the following performance concepts. The performance of any

Table 1.1 Basic Textile Definitions

Textile	An umbrella term for anything that can be made from a fiber or fabric. This is a very general term that could refer to a tennis ball cover, a disposable diaper, a dryer sheet, geotextiles (building materials), carpeting, or interior and apparel fabrics.
Fiber	The smallest visible part of a textile and the raw material of a fabric. A fiber is a hairlike strand very similar to your own hair. Fibers can be staple or filament and made of natural or manufactured materials.
Yarn	A number of fibers that are twisted or laid together to form a continuous strand. In order to make fabric, short fibers must first be made into longer, more usable lengths called yarn. Yarns can be spun or filament. Historically, figuring out how to do this took humankind a very long time.
Fabric	A method of construction or an organization of fibers and yarns. The most common fabric constructions are weaves and knits, but there are other fabrications as well. End use products (such as drapes) are made from fabrics.
Dyeing	The science of applying color to textiles.
Printing	The process of applying color in a design to textiles.
Finishes	Any process that is done to a fiber, yarn, or fabric to change the way it looks, feels, or performs. A fabric can be dramatically changed from its original appearance or performance by the way it is finished.

Table 1.2 Classification of Fibers

Protein fibers come from an animal and include all hair and fur fibers, such as wool, cashmere, angora, mohair, camel, and silk.
Cellulose fibers come from plant fibers. Cotton and flax are the most common, but others include ramie, sisal, jute, coconut, and corn.
Manufactured fibers are manufactured from natural fibers, but are altered from their original form, often by being melted down with chemicals. These fibers perform much like the chemical compounds from which they are derived (protein or cellulose), and include fibers such as rayon, acetate, modal, lyocell, and bamboo.
Synthetic fibers are purely manufactured without cellulose or protein, including polyester, acrylic, nylon, olefin, aramid, and spandex.
Mineral fibers are made from minerals, including gold, silver, and steel.

given textile is determined by the properties of the fiber, yarn, and fabric; every component inherently has these properties. In the chapters to come, these performance concepts will be extended to all the components of fabric construction. In this way, you will learn to make good choices that perform beautifully.

Textile Performance Concepts and Properties

Performance concepts relate to the measure of a textile's ability to perform in the final product. These concepts include durability, comfort, care, appearance, and safety.

Properties of Durability

Durability is the measure of a textile product's ability to resist stress and serve its intended use. (Each criterion can be measured in a textile lab.)

- **Abrasion resistance:** The ability of a fabric to withstand rubbing without wearing a hole in the surface. This property is relevant for upholstery, towels, bedding, and carpeting. Abrasion is measured by a Martindale or Wiezenbeck rubbing machine that measures the number of rubs a fabric can withstand before it loses fiber or wears a hole in the fabric. Often these numbers are listed on sample cards as evidence of the expected performance for its intended upholstered use. A consideration of placement is warranted. Is this for an occasional chair or a hard working arm chair for the family room?
- **Pilling or Shedding:** The formation of tangled fibers on the surface of the fabric. Pilling is also caused by rubbing but with a different end result.
- **Feltability:** The ability of fibers to matte together to form a fabric. Carpet underlay is often made of matted fibers.
- **Elasticity:** The ability of a fiber to stretch and recover to its original size and shape. A fiber with elasticity prevents sagging and bagginess in tightly upholstered pieces. This property is different from simple stretchiness.
- **Elastomericity:** The ability of a fiber to stretch 100 percent and recover. Consider spandex and rubber.
- **Elongation:** The degree to which a fiber may be stretched without breaking; the amount of give in a fabric. Growth can be a problem in textiles that simply elongate and do not recover from stretching.
- **Dimensional stability:** The ability of a fiber to retain a given size and shape. Relates to shrinkage.
- **Tenacity/tensile strength:** The ability of a fabric/fiber to withstand a pulling force. Relates to the breaking point.
- **Tearing strength:** The ability of a fabric to resist tearing or ripping.
- **Snag resistance:** The ability of a fabric to resist yarns being pulled out.
- **Yarn distortion:** The ability of a fabric to resist yarn or seam slippage. This is relevant to tightly upholstered products.
- **Flexibility:** The ability of a fiber to bend repeatedly without breaking. The more flexible the fiber, the more drapable the fabric.

Properties of Comfort

The following traits measure a textile's ability to support human comfort. This includes issues of static and temperature. Comfort does not relate to hand or the way a fabric feels.

- **Hydrophilic fibers:** Hydrophilic fibers are absorbent. This is an important quality that allows the fiber to absorb perspiration from the body for human comfort. Relevant for bedding and towels. Absorbency also assists in cleaning and stain removal.
- **Hydrophobic fibers:** These fibers absorb very little moisture. Textiles that are hydrophobic dry very quickly. They also resist water-based stains. Once stained, they can be difficult to clean because they do not absorb water for cleaning purposes. Absorbency is important as it impacts skin comfort, static, washability, shrinkage, wrinkling, and stain removal. The more hydrophilic the fiber, the more it shrinks, wrinkles,

and stains, but it has less static. Hydrophobic fibers shrink less (or not at all), wrinkle less, absorb fewer stains, but they can be quite static-y.

- **Hygroscopic fibers:** The ability of a fiber to absorb water vapor (moisture from the air) without feeling damp.
- **Wicking:** The ability of a fiber/fabric to pass moisture (perspiration) to the surface for evaporation.
- **Electrical conductor:** A fiber/fabric with the ability to transfer electrical charges. Electrical conductivity means no static.
- **Electrical retainer:** A fiber/fabric with the ability to build up electrical charges. Electrical retention means the fabric will have static.
- **Thermal conductor:** A fiber/fabric with the ability to transmit heat from the body. Relevant for bedding.
- **Thermal retainer:** A fiber/fabric with the ability to hold heat next to the body. Thermal retaining fabrics are warmer. Good for blankets.
- **Allergic potential:** The ability of a fiber/fabric to cause a physical reaction, such as skin irritation or watery eyes.
- **Weight:** A lightweight fiber enables a fabric to be warm, without being heavy, and can be made into thicker, fluffier fabrics without adding weight.
- **Air permeability:** The ease with which air passes through the fabric.
- **Cover:** The ability of a fiber to occupy an area.

Properties of Care

Care is the treatment that is required to maintain the original look of a textile product; it relates to wear, cleaning, and storage.

- **Resiliency:** The ability of a fiber/fabric to return to its original shape after bending, compressing, twisting, or crushing. Relates to wrinkle recovery.
- **Oil affinity:**
  - *Oleophilic* (oilophilic): means "oil loving"; holds oily stains. Must be cleaned with dry cleaning agents or enzyme cleaners.
  - *Oleophobic* (oilophobic): means "oil hating"; resists oily stains.
- **Chemical reactivity:** The ability to interact on a molecular level with cleaning agents, such as acids, alkali (bases), oxidizing agents, solvents, enzymes, bleaches, or others. This impacts the ability to use particular chemicals for cleaning.
  - Harmed by acids means that fruit stains should be removed immediately to prevent setting.
  - Resistant to alkali means that detergents can be safely used.
  - Harmed by alkali means that a mild or neutral detergent should be used. It also means that the textile will be harmed by perspiration.
  - Harmed by oxidizing agents means that chlorine bleach should not be used and that sunlight will cause whites to yellow.
- **Chemical benefits:** Some new fibers on the market contain trace minerals, vitamins, amino acids, antioxidants, and anti-inflammatory agents that can be absorbed transdermally.
- **Environmental effects:**
  - *Biological resistance:* The ability (or lack thereof) to resist moths (wools), mildew and silverfish (cellulose), and so on. This property relates to storage considerations. Caution should be used in thoroughly drying fabrics with poor biological resistance.
  - *Sunlight resistance:* The ability to withstand degradation from direct sunlight. This does not describe fading.
  - *Aging resistance:* The ability to maintain original qualities over time, which relates to storage and longevity.
  - *Biodegradability:* The ability to decompose naturally.
  - *Fume fading:* Color loss due to environmental gases or pollution.
- **Thermoplasticity:** The ability to soften, melt, or shrink when exposed to heat. This aids in the production of permanent pleats, creases, or embossed shapes. Thermoplastic fibers are used to create a textured filament that adds bulk and crimp in carpets. The fiber becomes pliable and moldable when heated and will retain a predetermined shape when cooled (heat set).
  - *Heat tolerance:* The ability to withstand heat when subjected to cleaning, drying, and ironing. Can iron at any temperature.
  - *Heat sensitivity:* The temperature at which a fiber will melt, soften, or shrink when subjected to heat. Relates to cleaning, drying, and ironing. Fabric may melt with too hot an iron. If harmed by heat, use steam; wool turns brittle and harsh and—will scorch when subjected to high, dry heat.

Cleaning codes

As a general rule, the lifespan and appearance of interior fabrics is increased by frequent vacuuming or brushing, whether it is for carpeting, upholstery or drapes. When ordering fabric for a specific use from most design houses, fabrics are usually marked with standardized codes for ease in identifying care. Pretesting on small, unobtrusive areas is always recommended. The following are the most common codes.

Code W—Water based cleaning. Spot clean as necessary using foam from a water-based cleaning agent such as a mild detergent or non-solvent upholstery shampoo. Follow instructions for application. Vacuum when dry.

Code S—Solvent or Dry Cleaning. Spot clean using a mild, water-free solvent or dry cleaning product. Follow instructions carefully and clean only in a well ventilated room.

Code W-S—Water or Solvent (Dry) Cleaning. Spot clean using a mild solvent, an upholstery shampoo or the foam from a mild detergent. Follow instructions carefully.

Code X—Vacuum or brush only. Do not use water or solvent based cleaners, due to the potential for shrinkage, distortion of the surface pile, or staining.

Properties of Appearance

Appearance is the measure of a fabric's ability to remain aesthetically acceptable and serviceable.

- **Shape and length of fiber:** The characteristics that influence appearance, care, and comfort. A short (staple) fiber performs differently from a long one (filament).
- **Color:** The first aesthetic property noticed in a textile product.
- **Colorfastness:** The ability of a fiber/fabric to retain color when exposed to environmental or chemical conditions, such as bleach or sunlight.
- **Dye affinity:** The receptivity of individual fibers to coloration by a particular dyestuff. Dependent on fiber base and absorbency. There is no single dye that dyes all fibers equally.
- **Luster:** The sheen created by light reflection. Filament fibers have more luster than staple.
- **Translucence:** The ability to pass light through a fabric.
- **Drape:** The manner in which a fabric falls or hangs. Created by flexibility.
- **Body:** The stiffness, or firmness, of a fabric. The opposite of drape.
- **Texture:** The three-dimensional nature of the fiber, yarn, or fabric surface.
- **Loft:** The ability of a fiber or fabric to spring back to original thickness after being compressed (compressional resiliency). (Does a carpet leave your footprints in their wake, or do they spring back up?)
- **Hand:** This is a description of the way a fiber feels to the touch: soft, smooth, crisp, dry, textured, and so on. People shop with their hands; a pleasing hand is critical.

Properties of Safety

The safety of a fabric for specific end uses is determined by how a fiber burns. This includes bedding, drapery, carpet, and upholstery. If a product in the home, such as a sofa, is quick to burn (easily combustible), there is potentially great risk for human life. A slower, smaller burn provides the necessary time for escape. The Flammable Fabrics Act Amendment of 1967 prohibits the sale of highly flammable fibers or fabrics for drapery, carpets, bedding, and upholstery. Since 1972, oversight and testing for these regulations have been provided by the Consumer Product Safety Commission (CPSC). Some of the considerations are as follows:

1. Does it simply burn or does the fiber support combustion (add fuel to the fire)?
2. How fast does it burn? If it burns too quickly, chances of getting out of the burning structure diminish.
3. How much smoke and toxic fumes are emitted?

Contract (industrial) fabrics have much higher standards for testing and burning than residential.

Aside from life-saving considerations, the burn properties aid in the identification of the fiber. (See Tables 4.4 and 4.5 of Chapter 4.) In general, there are four categories of burn characteristics:

- **Inflammable** (also known as flammable): Ignites and burns easily, with a bright orange glow. Inflammable fibers smell like burning paper or wood and do not easily extinguish; cellulose fibers are inflammable.
- **Flame retardant:** Fibers burn but self-extinguish when removed from flame. Flame-retardant fibers smell like burning hair; protein fibers are flame retardant.
- **Flame resistant:** Fibers burn and melt. Flame-resistant fibers often self-extinguish when removed from flame. The fibers smell like burning plastic; synthetic fibers are flame resistant.
- **Nonflammable:** Fibers do not burn. Flameproof fibers/fabrics will not ignite; mineral fibers are nonflammable.

Activity 1.1 Research Project: New Textiles

Name: ____ Date: __________

Research websites, magazines, or industry publications to find an article about an exciting new textile; look for new technological advances, such as those referenced in this chapter. There are many more examples of new technology in the textile field. Because you are looking for a truly new innovation, the article should be no more than six months old.

Write a summary of your article; be sure to mention what is new and different about the textile. Also include why it is exciting to you personally and how you, as a designer, could use this new textile in an interior environment. Be creative!

In selecting an article, be sure to avoid forecasting and trend articles, as well as public relations pieces by large corporations (who want you to believe that everything that they do is new and exciting). Also be certain that the textile you select is new. Include a copy of the article with your summary, along with the source and date.

CHAPTER TWO

Fiber Classification: Natural Fibers

Reference Swatches

5. Cotton | 6. Organically Colored Cotton | 7. Flax
8. Hemp | 9. Jute | 10. Raffia | 11. Wool | 12. Mohair | 13. Silk
14. Wild Silk | 15. Silk Noil | 16. Dupioni Silk

Overview: Natural and Manufactured Fibers

Critical to the comprehensive understanding of textiles is the chemical base of each fiber. All fibers can be grouped into one of the following four chemical classifications:

- Cellulose: Fibers that come from a plant, such as cotton.
- Protein: Fibers that come from an animal, such as wool.
- Mineral: Fibers that come from minerals, such as glass.
- Synthetic: Purely chemical fibers, such as nylon.

By learning the chemical base of a fiber, it becomes much easier to understand and anticipate why fibers perform in certain predictable ways. Table 1.2 (page 6) and Figure 2.1 (page 17) offer quick references that distinguish the base of the fiber.

Cellulose Fibers

Swatches 5–10 are made of cellulose fibers. Cellulose is any fiber that comes from a plant. As you are looking at the fabric swatches, remember that fiber is simply the hairlike strand that makes up the yarn, which is then organized into a fabric structure. Notice that cotton fabrics can be quite different from one another. Check each fabric for hand, but remember that fabric weight and structure play a large part in differentiating one fabric from another.

Seed Fibers

Cotton

Appreciated for its absorbency (hydrophilic), breathability (ability of a fiber to transmit air and moisture), and soft hand, cotton has been the fabric of our lives for more than 7,000 years. A staple in the textile world, this versatile fiber is equally appropriate for bedding, towels, window treatments, infant's wear, shirts, pants, dresses, canvas bags, and lingerie. Cotton is the most important fiber for sheets due to its hydrophilic nature, which means that it absorbs moisture (perspiration) from the body, leaving the sleeper comfortable and dry. The many faces of cotton are evident in this text, with many different examples, including fabrics such as denim, jersey, terry, embroidered eyelet, and many more.

Cotton is a particularly short staple fiber (ranging 0.5–2.5 inches), comprising about 31 percent of the world's fiber consumption, predominantly produced in China, India, and the United States. The fiber comes from the boll or seed of the plant. Once harvested, cotton is graded as a commodity according to color, staple length, and cleanliness. White cotton is the highest grade (provides the truest dyed color), as are the longer staple lengths. Cleanliness relates to the amount of leaves and stems, called trash, left in the fiber. Better quality cottons are longer staple fibers and include Sea Island, Pima, and Egyptian varieties. The longer fibers are silkier, stronger, and more lustrous than regular Upland or Alcala cotton. Table 2.2 provides an extended list of performance properties for cotton.

Colorgrown cotton means that the cotton is actually grown in color—no need to dye it. Cotton has a recessive gene for color and can be grown naturally in a variety of soft colors, such as greens, browns, and rusts. As it is not dyed, it leaves less of a footprint on the environment and does not fade, bleed, or crock.

Although cotton is biodegradable, it is only so if it is not left in landfills. Cotton and other natural fibers are considered sustainable, as they are renewable. Traditional cotton requires a lot of water and pesticides to get it to harvest, however. In fact, cotton is chemically dependent; it can require as much as one-third a pound of pesticides for each T-shirt made and uses about 3 percent of the world's water supply. Organic cotton means that the fiber is grown with a low impact on the environment, without the use of toxic fertilizers and pesticides. To be certified organic, the field in which it grows must be pesticide-free for three years. Often when a label suggests organic cotton, it really means that a portion of the cotton used in the product is, in fact, organic, as only 2 percent of the world's cotton is organic. New labels are being created that will inform the consumer in which field or farm their fiber was grown.

Cotton-rich is a relatively new term (often applied to sheets) that means that the fabric is made of at least 60 percent cotton; the remainder is often polyester.

Transitional cotton means pesticides are no longer being used in the production of this cotton. This is an easier requirement for the farmer than organic cotton, releasing them from the three years without pesticides commitment.

Recycled cotton has historically been the lowest quality, as it has traditionally meant shredding the fibers down into shorter fibers to re-spin. This provides a relatively rougher hand.

Used clothing and clipping waste from garment construction and loom waste are now being harvested and upcycled into longer staple fibers for production. Postconsumer textiles (PCT) are finding useful new lives and can be a viable, low-impact, and environmentally friendly alternative for recycled cotton.

[Reference Swatches 4–6]

Bast Fibers

Flax

Flax is a bast fiber, meaning the fiber is harvested from the stem of the plant. It is therefore a much longer staple than cotton (6–24 inches). In common use for about 10,000–12,000 years (e.g., mummy wrappings), fabrics made from flax or other bast fibers are usually called linen. Flax is imported from colder European climates, such as Belgium, Ireland, France, and Poland.

Chosen for its texture, flax has an irregular surface caused by the natural unevenness of the stem of the plant. Three times stronger than cotton, flax softens with each wash, is more absorbent than cotton, and dries quickly. This is why flax is an excellent choice for warm weather. On the other hand, flax fibers have very poor resilience (wrinkle recovery). Choose this fiber for body, not drape; the most outstanding characteristic is its surface texture. Table 2.2 provides an extended list of performance properties for flax.

[Reference Swatch 7]

Ramie

Although ramie was used in prehistoric China, Europe, and Egypt, this bast fiber was not commonly used until the past decade due to the difficulty in processing. Recent technological advances have made ramie more accessible for production. Stronger and more absorbent than flax and cotton, it takes dyes better, is quick drying, and does not mildew or rot. The fiber has a slightly lustrous appearance that improves with washing, and it has better dimensional stability than the other cellulose fibers.

Hemp

The third of the bast fibers, hemp has been in production for 10,000 years. This strong fiber looks similar to flax and, according to the hemp lobby, is the world's most versatile fiber with 20,000 documented uses. Hemp is recognized for its environmentally friendly image because it grows so quickly without the use of pesticides. The fiber can make paper quickly, without deforesting the world. It will also provide us with furniture, upholstery, clothing, plastics, methanol (clean fuel for our cars), a protein-rich food, shampoos, lotions, detergents, paint, and, of course, medication. And if that is not enough, the fiber is naturally resistant to mold, mildew, and rot, a traditional problem with most cellulosic fibers. Even if only half of its proponent's claims are true, hemp is still the most useful agricultural crop in the world.

Tables 2.1 and 2.2 offer quick references for fibers, their chemical bases, and their specific properties, along with their significance for the consumer.

Jute

Jute is a relatively coarse fiber that lends itself well to making twine, carpet backings, rugs, rope, chair coverings, and flour and seed sacks. It has good insulation qualities and is not prone to static.

[Reference Swatches 8–9]

Leaf Fibers

Sisal

Sisal is a leaf fiber that comes from the agave or yucca plant and is a staple of welcome mats and area rugs. The fiber does not hold dirt, is static resistant, and offers sound and absorbing qualities. It can also be treated with a flame retardant chemical. Sisal weathers well but is not colorfast to sunlight. Sisal is often selected for outdoor mats.

Raffia

Raffia is stripped palm leaves and a good choice for outdoor furniture and rugs as it is a cellulose fiber that does not shrink with moisture.

[Reference Swatch 10]

Other Cellulose Fibers

Other cellulosic fibers such as coir, abaca (banana), and kapok have experienced reduced popularity.

Protein Fibers: Wool and Silk

Swatches 11–16 are made of protein fibers, which come from animals or insects. Table 2.3 provides specific types of wool. Tables 2.4 through 2.6 provide properties for protein fibers.

[Reference Swatches 11–16]

Wool

Wool is often used as a generic term to include the hair from any animal, but it generally references the hair of a sheep unless otherwise stated. In early history, skins were used. About 5,000 years ago, humans figured out that if we just sheared the sheep, it would be kind enough to grow us more fiber. Another staple, the fiber can be as long as 18 inches in length with a natural crimp. The fiber is greasy (like your own hair), in order to protect itself. The grease is called lanolin and is the softening substance of lotions and conditioners.

Under a microscope, the scales or cuticle of the fiber is evident. The scales are responsible for the fiber's propensity to shrink and felt and, sometimes, to feel itchy or scratchy in poorer grades. The cuticle is also responsible for the fiber's amazing abrasion resistance, which makes it a great candidate for carpeting. Superwash wools are coated with resin to smooth down the edges of the scales, which makes the fiber softer and prevents or minimizes shrinking and felting, creating washable wool. The natural spiral crimp makes wool springy and gives the fiber superior elasticity (stretch and recovery).

Wool is highly underrated as a fiber and the most misunderstood; it is equally appropriate for the ski slopes of Colorado and glitzy red carpet events, yet it enjoys only 1.5 percent of world consumption. Contrary to popular misconception, wool has wonderful drape, elasticity, and resilience. The fiber is an excellent choice for upholstery fabrics and carpet alike, due to its superior abrasion resistance.

Sheared wool fiber is called fleece and is graded in the following four categories:

1. *Color* (lighter is more desirable, as it is easier to dye). Wool cannot be bleached.
2. *Staple length* (longer is more lustrous and more valuable).
3. *Time of cut*
    a. The first cut is called lambswool and is softer and finer.
    b. Subsequent cuts are called fleece.
    c. The final cut is the lowest quality and is called pulled wool.
4. *Fineness of the wool* (relates to the cuticle and the breed).
    a. Merino and cashmere are among the finest.

There are more than forty different breeds of sheep, some with coarse hair and some with fine hair, just like humans. As a consumer, look for merino, cashmere, or lambswool (first cut) for very fine, soft wools. Novelty wools include mohair (from an angora goat), angora (angora rabbit), and the camelid family: camels, alpaca, llama, and vicuña. For a more extensive discussion of the properties of wool, see Tables 2.4, 2.5, and 2.6.

[Reference Swatches 11–12]

Silk

Silk is the only natural filament fiber, which is why it is the most lustrous and drapey. Legend has it that silk was discovered in 2600 BC by a Chinese princess when a silk cocoon fell into her hot tea and began to unravel. It is a logical enough story—perhaps even true. What is absolutely true is that the Chinese kept this secret for nearly 3000 years. Europeans believed that the Chinese were hiding herds of tiny sheep. In fact, the lustrous fiber comes from the cocoon of a silk caterpillar. The caterpillar eats only mulberry leaves and converts the leaves into the shiny, white silk fiber used to make its cocoon. The fiber is coated with a substance called sericin, which acts as the glue to hold the cocoon together. The fiber is very tiny; more than 1,500 cocoons are required to make a single yard of fabric.

Today we get most of our silk from China, India, Japan, Thailand, and Italy, and world consumption is still only about 1 percent. This is a luxury fiber, carefully cultivated into fine silk fabrics. The business of raising silkworms for fiber is called sericulture.

The industry offers two broad categories of silk: cultivated silk is fine filament silk, sometimes called mulberry silk (Bombyx Mori). Wild silk comes from shorter staple silk, broken cocoons, and Assam silk (Muga, Eri, and Pat silk).

Dupioni silk is harvested from two cocoons that grow together. It still has the luxury of fine filament silk, with a lovely uneven texture and nice luster.

Silk noil is made from waste silk; it is staple, with a very cotton-like hand. For a complete discussion of properties, see Tables 2.4, 2.5, and 2.6.

[Reference Swatches 13–16]

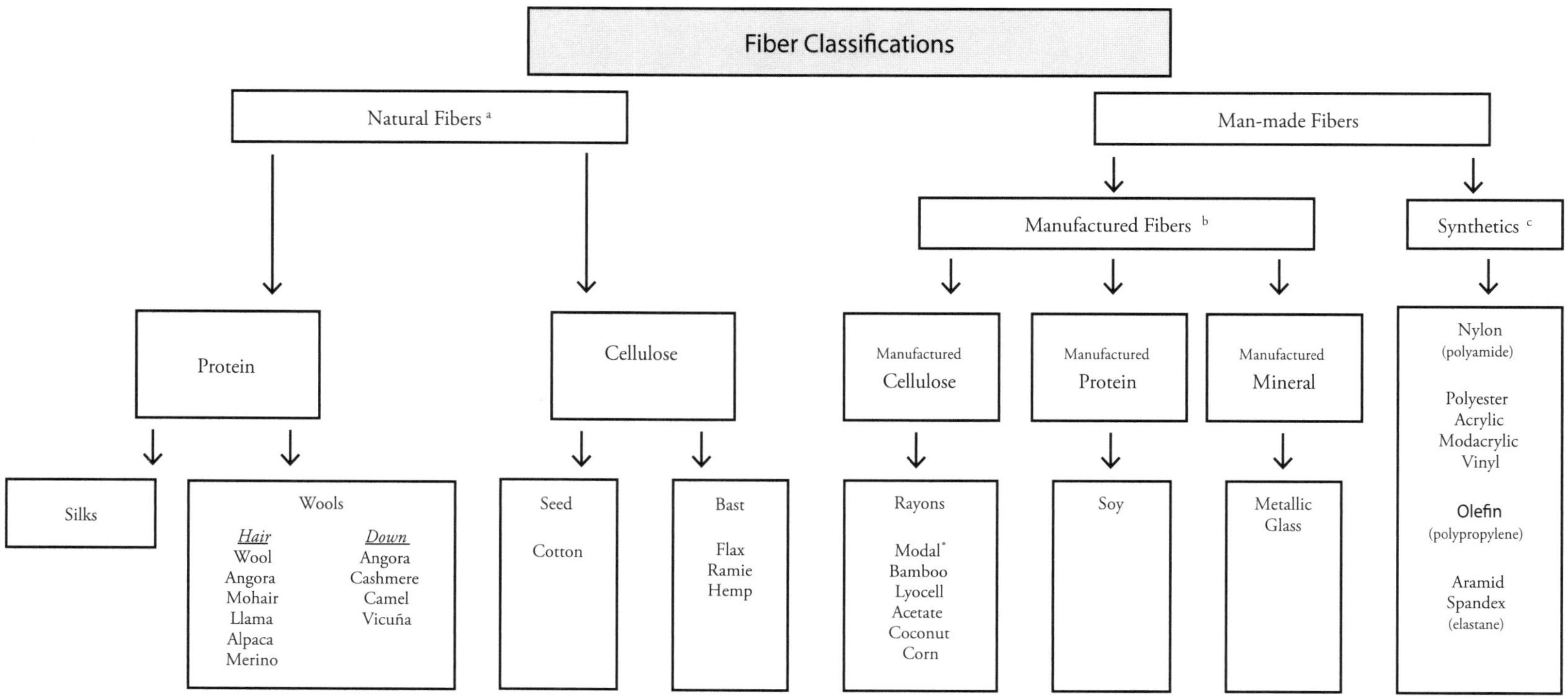

[a] Natural fibers are used in their natural state.
[b] Manufactured fibers can be made of natural ingredients, but can be processed with chemicals and used in an altered state.
[c] Synthetic fibers are purely chemical.

Figure 2.1 Fiber Classification

Table 2.1 Properties Common to All Cellulose Fibers: Cotton, Linen, Ramie, Hemp
(These properties also apply to the manufactured cellulose fibers: rayon, lyocell, bamboo, Modal®, acetate.)

Properties	Importance to Consumer
ADVANTAGES	
Hydrophilic (good absorbency)	Absorbs perspiration quickly; ease in cleaning. Natural fibers only (cotton, linen, ramie, hemp) are stronger when wet and can be washed. Manufactured cellulose fibers (rayons, lyocell, bamboo, Modal®, acetate) are weaker when wet.
Electrical conductivity	Conducts electricity away; does not build up static.
Thermal conductivity	Conducts heat away; cool on the body.
Heat tolerance	High temperatures are acceptable. No special precautions in ironing.
Chemical reactivity: Resistance to alkali (bases)	Can be laundered safely in detergents.
Pilling	None; natural cellulose fibers do not form matted fiber into pills. (Rayon can pill, however.)
DISADVANTAGES	
Resiliency	Wrinkles badly unless finished.
Elasticity	Does not recover from stretching.
Dimensional stability	Tends to shrink.
Drape	Natural cellulose does not drape well. Manufactured celluloses drape well.
Chemical reactivity	Harmed by acids; fruit stains should be removed immediately to prevent setting.
Environmental effects: Biological resistance	May be attacked by mildew (fungi) and silverfish. Textiles should not be put away damp.
Inflammability	Ignites quickly, burns freely, with an afterglow and gray, feathery ash. Filmy or loosely constructed garments should not be worn near an open flame.
NEUTRAL	
Strength and abrasion resistance	In consideration of all fibers, cellulose fibers rank medium to medium-high.
Low luster	Very little light is reflected off the surface of these fibers. Manufactured cellulose can be lustrous.

Table 2.2 Cellulose Properties Study Guide

Cellulose	Neutral Properties	Advantages	Disadvantages
Cotton	Strength Abrasion resistance	Hydrophilic (7%–11% moisture regain) and stronger when wet Electrical conductivity Thermal conductivity Soft hand No pilling	Drape Loft Resiliency Dimensional stability Environmental effects: Mildews and susceptible to silverfish Elasticity Luster Inflammability
Linen	Abrasion resistance	Body Texture Hydrophilic (12% moisture regain) and stronger when wet Electrical conductivity Thermal conductivity Tenacity No pilling	Drape Loft Extremely poor resilience Environmental effects: Mildews and susceptible to silverfish Flexibility Elasticity Dimensional stability Luster Inflammability
Ramie		Body Environmental effects: Mildew and mold resistant Texture Hydrophilic and stronger when wet Electrical conductivity Thermal conductivity Tenacity	Drape Loft Resiliency Flexibility Elasticity Dimensional stability Environmental effects: Susceptible to silverfish Inflammability
Hemp		Environmental effects: Mildew and mold resistant Hydrophilic and stronger when wet Electrical conductivity Thermal conductivity Tenacity No pilling	Flexibility Elasticity Dimensional stability Drape Loft Resiliency Luster Environmental effects: Susceptible to silverfish Inflammability

Note: Fabric weight, construction, or finish may change the performance of the fiber.

Table 2.3 Minor Hair Fibers

CAMEL FAMILY	
Alpaca	Soft, silky, highly insulating, lustrous, expensive; hair grows slowly.
Camel	Naturally water repellent, thermal retainer: excellent for coats; strong, smooth, lustrous.
Llama	Highly insulating, lightweight, resilient, colorfast, durable.
Vicuña	Finest fiber: strong, elastic, resilient; endangered species.
GOAT FAMILY	
Mohair (Angora goat)	Smooth, strong, reduced crimp,[a] washable, resilient; dyes well; does not felt or pill.
Cashmere	Very fine fiber with fine scales[b] or cuticle; soft, lightweight; not strong or durable.

[a]Crimp relates to cohesiveness.
[b]Scales are the cuticle of the fiber, which can sometimes cause itchiness or irritation.

Table 2.4 Properties Common to All Protein Fibers

Hydrophilic (good absorbency)	Comfortable in cool, damp climate; moisture prevents brittleness in carpets; hydrophilic quality creates ease in stain removal. Generally not considered washable, since weaker when wet. (Wool loses about 40% strength, and silk loses 15%).
Electrical conductivity	Conducts electricity away from the body; does not build up static.
Dimensional stability	Poor; subject to shrinking during laundry (wool substantially more so than silk).
Thermal retention	Retains body heat; warm as a blanket.
Weight	Feels lighter than cellulosics of the same thickness.
Chemical reactivity	Harmed by alkalies: can be washed in neutral or slightly alkaline soap or detergent. Harmed by oxidizing agents: chlorine bleach damages fiber and should not be used; use oxygen bleaches only.
Heat sensitivity	Wool becomes harsh and brittle and scorches easily with dry heat: use steam. White silk and wool turn yellow with too much heat. No steam on silk, as water produces spots.
Environmental effects: Biological resistance	Moth control needed for wool; carpet beetles eat silk.
Flame retardant	Does not burn readily; self-extinguishing; has odor of burning hair; forms a black, crushable bead.
Drape	Both flexible and drapable.

Table 2.5 Properties of Individual Protein Fibers

Protein	Advantages	Disadvantages
Wool	Texture Loft Resiliency Hydrophilic (13%–18% moisture regain) and weaker when wet Electrical conductivity Thermal retention Elasticity Abrasion resistance Flame retardant Drape	Dimensional stability Chemical reactivity: Bleach Environmental effects: Biological resistance (moths) Tenacity: Weakest natural fiber Pills No luster
Silk	Hydrophilic (11% moisture regain) and weaker when wet Electrical conductivity Thermal retention Tenacity: Strongest natural fiber Flame retardant Drape Luster Does not pill	Dimensional stability Chemical reactivity: Bleach Resiliency Elasticity Environmental effects: Sunlight resistance Biological resistance (carpet beetles) Abrasion resistance Perspiration resistance

Table 2.6 Comparison of Protein Fiber Properties

	Silk	Wool
Resiliency	Average to poor	Good: resists wrinkling; wrinkles hang out between wearings; fabrics tend to hold their shape
Elasticity	Poor	Good: Recovers well from stretching
Abrasion resistance	Poor	Excellent: A great choice for carpeting
Environmental effects: Perspiration	Poor perspiration resistance; perspiration weakens the fiber	Good perspiration resistance
Biological resistance	Carpet beetles can be a problem	Moths can be a problem
Environmental effects: Sunlight resistance	Poor: Sunlight damages the fiber. Silk drapes must be lined. Sunlight causes white silk fabrics to turn yellowish	Good sunlight resistance
Hygroscopic	Not hygroscopic	Hygroscopic
Luster	Has luster	Has no luster
Pilling	Does not pill	Pills
Fiber Length	Filament	Staple
Tenacity	Strongest natural fiber	Weakest natural fiber

Activity 2.1 Swatch Page: The Diversity of Cotton

Name: ______________________________ Date: __________

Swatch page: The Diversity of Cotton

Visit a fabric store or thrift shop and find six cotton fabrics that are very different from one another. The differences should encompass weight and structure, not just color or print. Start with fabrics that you know, such as denim and corduroy. Mount the fabrics on the swatch page provided and identify each swatch by fiber and fabric name.

Swatches should be cut to 2 × 3 inches and neatly presented (no hems or seams). Do not share swatches with another student, and do not use swatch book swatches.

Fabric name: ______________________________
Fiber content: ______________________________
End use: ______________________________

Fabric name: ______________________________
Fiber content: ______________________________
End use: ______________________________

Activity 2.1 Swatch Page: The Diversity of Cotton *(continued)*

Name: ______________________________ Date: __________

Swatch page: The Diversity of Cotton

Fabric name: ____________________
Fiber content: ____________________
End use: ____________________

Fabric name: ____________________
Fiber content: ____________________
End use: ____________________

Fabric name: ____________________
Fiber content: ____________________
End use: ____________________

Fabric name: ____________________
Fiber content: ____________________
End use: ____________________

CHAPTER THREE

Fiber Classification: Manufactured Fibers

Reference Swatches

17. Rayon | 18. Bamboo | 19. Micromodal® | 20. Acetate
21. Ingeo® Corn | 22. Glass | 23. Metallic

Manufactured fibers include two broad categories of fibers: man-made and synthetic. Man-made fibers consist of natural materials (protein or cellulose fibers) that have been altered (processed) with chemicals. These fibers are not used in their original state (as found in nature); they are melted down with harsh chemicals and extruded from a spinneret into long filaments. The spinneret looks like a shower head, with liquid fibers that harden once extruded. Man-made fibers often perform more like natural fibers than their synthetic counterparts, due to their natural ingredients. The second category is synthetic fibers, which are totally chemical in origin.

Manufactured Cellulose

Manufactured cellulose fibers are made from plant material that are altered from their original form, and processed with chemicals. This processing causes a permanent change in the structure of the fiber. For this reason, although the fibers are made from cellulose, and perform like cellulose, they are classified as manufactured fibers. For a list of properties appropriate to all manufactured cellulose fibers, see Table 2.1, Properties Common to All Cellulose Fibers (page 18).

Rayon

In an attempt to replicate silk, the development of rayon (1890s) was the first successful foray into fiber production. Taking our cue from silk caterpillars, mulberry leaves (the silkworm's natural diet) were melted down and extruded through a spinneret to create the fiber. For this reason, rayon was dubbed artificial silk.

Mulberry leaves are cellulose, thereby giving the fiber many cellulose attributes (although it does have many silk qualities). Today the fiber is made of many different trees. Rayon is a thermal conductor (transfers body heat), hydrophilic (absorbent), and wicks well (transfers moisture to the surface for evaporation).

As a result of its manufacture, the fiber is a filament and is lustrous with excellent drape. The draping quality, along with an incredibly soft hand, is what makes it so seductive in today's market. Use rayon for an interior drape (it has poor sunlight resistance), pillows, or other accessories but not for upholstery, due to its compromised strength.

Rayon is the generic name and is also known as viscose (particularly in Europe). Since its creation, there have been several innovations in this fiber line. Manufactured cellulose today comprises 6 percent of world fiber consumption. High wet modulus rayon (HWM) has an improved wet strength. Cuprammonium Rayon (often called cupro) is a finer, improved variant that is more silk-like with a very distinctive hand. Bemberg® is the only tradename for this fiber. See Table 3.1 for distinctive properties.

[Reference Swatch 17]

Table 3.1 Properties of Individual Manufactured Fibers

Manufactured Cellulose	Advantages	Disadvantages
RAYON Viscose Bemberg® cuprammonium (often called *cupro*) High-wet-modulus (HWM, or polynosic) Modal® Bamboo	Drapability Hydrophilic (13% moisture regain) Electrical conductivity Thermal conductivity Hand: Soft Wicking Antibacterial (Bamboo and Modal®)	Resiliency Dimensional stability Chemical reactivity Tenacity Elasticity Environmental effects: Mold, mildew
ACETATE Celanese	Hydrophilic (6% moisture regain) Hand: Smooth and silky Luster Drape	Chemical reactivity Tenacity Elasticity Resiliency Dimensional stability Environmental effects: Fume fading, mold, mildew Electrical retainer
LYOCELL Tencel®	Hydrophilic Electrical conductivity Thermal conductivity Tenacity Resiliency Dimensional stability Hand: Soft Drape	Elasticity Environmental effects: Biological mold and mildew
COCONA® (Coconut fibers) SORONA® (Corn fibers)	Hydrophilic Electrical conductivity Thermal conductivity Hand: Soft Resiliency Dimensional stability Wicking Antibacterial properties Sunlight resistance Biodegradability Cocona® has superior UV protection and absorbs odor	Flame resistance: Melts
SEATIVA (Seaweed and Lyocell)	Hydrophilic Electrical conductivity Thermal conductivity Chemical benefits: Full of trace minerals, vitamins, and amino acids Antioxidant properties Anti-inflammatory agent Hand: Soft Luster	*[Independent lab results pending]*

(Table continued on next page)

Table 3.1 Properties of Individual Manufactured Fibers (*continued*)

Manufactured Cellulose	Advantages	Disadvantages
SOY	Hydrophilic Electrical conductivity Thermal conductivity Resiliency Dimensional stability Wicking Antibacterial properties 18 amino acids absorbed transdermally	Elasticity Environmental effects: Biological resistance Tenacity: Blended for this reason *[Independent lab results pending]*
Manufactured Mineral	**Advantages**	**Disadvantages**
GLASS Fiberglass (not used in apparel)	Tenacity Nonflammable Chemical resistance	Irritation potential Abrasion resistance Elasticity: Poor Hydrophobic Heavy weight
METALLIC Aluminum	Luster	Strength: Poor Purely decorative

Bamboo

This fiber is made from the bamboo plant and is called bamboo rayon (or rayon from bamboo), distinguishing it from its predecessor (rayon). The fiber is well known for its amazingly soft hand, instant wicking ability, and antibacterial qualities, which make it a natural in the towels and sheets market. There is some question, however, as to whether the fiber retains its antibacterial qualities after the process of regenerating. The fiber has interesting odor-absorbing qualities and has found a home in the sock market, as well as in shirting and infant wear. Bamboo is cooler, lighter weight, more breathable, and softer than cotton. The front end of the process, growing the fiber, is very good for the environment. Bamboo is sustainable and grows quickly without pesticides. Additionally, it dyes beautifully and with significantly less dye.

After harvesting, however, harsh chemicals and a lot of energy are used to melt down the fiber to extrude from the spinneret. There are other alternatives in processing, but they are seldom used due to the expense.

[Reference Swatch 18]

Modal

Modal is a fine HWM rayon and is often created as a microfiber called Micromodal®. This fiber is made from beech trees and is well known for its very silky hand and exceptional drape. Often presented as a fine-gauge jersey knit, the fiber is 50 percent more absorbent (hydrophilic) than cotton, yet with reduced shrinkage (dimensional stability). Produced by Lenzing, this fiber is quickly becoming a staple in the contemporary fashion market and is finding its way into the towel and bedding market for interior design.

[Reference Swatch 19]

Lyocell

Lyocell (developed in 1993) is also a manufactured cellulosic fiber made from eucalyptus trees. This is an ecologically green fiber (closed loop system) made with recyclable organic solvents (instead of water). Lyocell is more absorbent (hydrophilic) than rayon, yet it shrinks less (dimensional stability) and wrinkles less (resilience). This is really a greener and more environmentally friendly version of rayon. Tencel®, produced by Lenzing, is the most common trade name for this fiber. See Table 3.1.

[Reference Swatch 69]

Acetate

Invented by the Celanese Company in 1924, acetate was a further attempt at simulating silk. Composed of cellulose fibers and chemicals, it is processed with acetone. For this reason, acetate is the first thermoplastic fiber. Look for it to burn and melt in the burn test in Chapter 4.

Acetate is commonly used in linings and special occasion gowns with limited use due to its poor qualities. The poor strength and abrasion resistance is often experienced in disintegrating linings of a favorite jacket or handbag. The outside or shell fabric remains intact, yet the lining is often shredded. This is the nature of acetate. The fiber has maintained its market share because it offers beauty (hand, luster, and drape) at a low cost. Acetate was once dubbed "the beauty fiber" for this reason. Acetate has found a market in accent pillows, satin or taffeta window treatments, and bedspreads. These are a good choice due to their limited contact.

[Reference Swatch 20]

Triacetate is a generic name of a modified acetate with more chemical than cellulose and, as such, can be heat set. Like acetate, it has a luxurious hand, great luster, and drape.

Cocona®

Cocona® is the trade name for the fiber made from coconut shells, which is essentially a waste product. It is usually blended with polyester and spandex. Cocona® fabric is known for superior wicking, moisture management, and UV protection. It is also a good choice for odor management, as odors get trapped inside of the activated carbon.

Manufactured Protein

Manufactured protein fibers are also made of natural ingredients and, once again, not considered natural fibers because they are processed with chemicals and used in a form different from their origins. The generic name for man-made proteins is Azlon.

For all practical purposes, this is the newest category, with some exciting new innovations. The properties for these fibers are mostly the same as the protein fibers. See Table 3.1.

[Reference Swatches 18–24]

Soy

Soy was first discovered by Henry Ford in 1937 when he wore a suit made of a fiber he called soy wool. In 1998, it was reinvented and is today made from tofu waste. Soy is a botanical protein (protein from plants). Often called Soysilk, soy cashmere, or vegetable cashmere, the fiber easily imitates both silk and wool. Soy latte is a new fiber made from thick soy milk proteins that are extruded into a fiber. Soy is quite green; since it uses the leftover dregs of soybean oil, it has no footprint of its own. The fiber is being produced as bedding, underwear, yoga wear, and socks due to its very soft and silky hand. The possible benefits of wearing soy relate to the same health advantages of ingesting it. When it was first produced, it was used alone, but experience has taught manufacturers that soy is best used in a blend, as it is not strong enough to be used alone.

Seaweed

Seaweed has been blended with lyocell in a fabric called SeaCell™; the potential benefits include the 18 amino acids that can be absorbed through the skin while wearing/using the fiber. As an antioxidant, it can also assist with skin irritations. This would be an excellent choice for sheets, but it is also soft and lovely for blouses and shirting. It has been marketed under the name of SeaTiva™, made of SeaCell™ and cotton.

Manufactured Mineral

Like the other fibers in this category, man-made mineral fibers are made of natural materials and processed with chemicals. These fibers do not tarnish and are used for decorative purposes.

Glass

Glass fibers are made from the silica in sand. It is the heaviest fiber; it is extremely hydrophobic and has no elasticity. In fact, if the fiber is repeatedly bent in the same place, it may break. Humans have a tendency to bend in the same places, and, for this reason, it is not a wearable fiber. Notice the softness and apparent flexibility of the fiber, however. It is used for nonflammable window treatments, filters, fiber optics, insulation, and, in this case, boating materials with which to add resins to the surface of the boat.

[Reference Swatch 22]

Metallics

Once made of actual gold and silver, metallic yarns today are essentially colored aluminum foil with a polyester coating. As defined by the Federal Trade Commission, "a metallic fiber is composed of metal, plastic-coated metal,

metal-coated plastic, or a core completely covered by metal." Commonly, this means that colored aluminum foil is coated with aluminized salts and is then cut into narrow strips. The resulting fiber is purely decorative due to its extremely poor strength.

Often using the trade name of Lurex, these fibers do not tarnish and are like Christmas tree tinsel: flashy and decorative but not sturdy. When used in fabric, the fiber is generally used only in the horizontal direction due to its lack of strength. Fabric made of mostly metallic yarns is called lamé.

[Reference Swatch 23]

Stainless Steel

Used extensively in carpets to reduce static, stainless steel, carbon, and other metals are being used to add strength and abrasion resistance. Looking forward, they will also be able to transmit and radiate heat when connected to a power source. For these reasons, metals will command a larger share of the interior design marketplace.

Mineralé

Cocona® also created a fiber called Mineralé that is made from volcanic minerals and will purportedly assist in evaporative cooling, odor resistance, and UV protection.

Activity 3.1 Swatch Page: The Diversity of Polyester

Name: ______________________ Date: __________

Swatch page theme: 1 fiber, 6 different fabrics

Visit a fabric store and find polyester fabrics that are very different from one another.

The differences should encompass weight and structure, not just color or print. Start with fabrics that you know, such as polyester chiffon and polyester brocade. Mount the fabrics on the swatch page provided and identify each swatch by fiber and fabric name.

Swatches should be cut to 2 × 3 inches and neatly presented (no hems or seams). Do not share swatches with another student, and do not use swatch book swatches.

Fabric name: ______________________
Fiber content: ______________________
End use: ______________________

Fabric name: ______________________
Fiber content: ______________________
End use: ______________________

Name: ______________________________ Date: __________

Swatch page theme: 1 fiber, 6 different fabrics

Fabric name: ______________________________

Fiber content: ______________________________

End use: ______________________________

Fabric name: ______________________________

Fiber content: ______________________________

End use: ______________________________

Fabric name: ______________________________

Fiber content: ______________________________

End use: ______________________________

Fabric name: ______________________________

Fiber content: ______________________________

End use: ______________________________

CHAPTER FOUR

Fiber Classification: Synthetic Fibers

Reference Swatches

24. Nylon | 25. Acrylic | 26. Polyester
27. Olefin | 28. Nomex® Aramid

Introduction to Synthetic Fibers

By 1939, the process of manufacturing fibers extended into using resources such as petroleum products, petrochemicals, natural gas, and coal. The raw materials undergo very complex processes necessary to spin the materials into fiber.

Historically, the reputation of these fibers has suffered, due in part to the legacy of the leisure suit of the 1970s. There is also a generally held belief that synthetic fibers don't breathe, have a poor hand, are static-y, and present environmental concerns. Synthetic fibers have come a long way since the 1970s and often compare favorably with the better-respected natural fibers. Today they command a respectable 61 percent of world consumption. A softer hand, wicking, breathability, and a social awareness of recycling and sustainability in their production are expected components of today's synthetic fibers. For these reasons, synthetic fibers have become dominant in many markets, including and especially the interior design markets. If you are looking for wash-and-wear, easy care, stain resistance, flame resistance, or general performance fibers, synthetics should be your first consideration. Moreover, synthetic fibers are often blended with natural fibers, so that consumers can enjoy the hand of natural fibers, while enjoying the benefits of synthetics. The blend may be orchestrated to improve strength, dimensional stability, and resilience to their natural counterparts. In effect, they make natural fibers better behaved. In upholstery, the hydrophobic nature of synthetic fibers render them stain resistant, not absorbing body moisture or water-based stains. For these and other reasons, the fibers in this category have been employed full-time since their original entrance into the marketplace. See Tables 4.1, 4.2, and 4.3 for an expanded list of properties attributed to all of the following fibers.

Nylon

Nylon, made of petroleum, was the first purely synthetic fiber created by DuPont in 1939. Originally dubbed fiber 66, it is now also called *polyamide*, particularly in Europe. Nylon is usually chosen for its light weight, natural transparency, incredible strength, and abrasion resistance. Conversely, nylon does have issues with static and pilling, but science has found ways to mitigate these drawbacks. Nylon is often blended with other fibers to add strength, as it is the strongest of the commonly used fibers. As such, it is widely used for luggage, backpacks, and parachutes (where strength is essential). Nylon is also the top choice for carpeting and upholstery due to its compressional resilience, strength, abrasion, and stain resistance. Astroturf is made of nylon, as is Velcro®. Due to its inherent transparency, nylon is often a good choice for sheers and lace structures.

Common trade names to look for are Supplex®, a textured nylon; known for its cotton-like hand and good wicking, it is lightweight and windproof. Tactel® is a stretch textured nylon, and Cordura® has excellent strength and abrasion resistance.

[Reference Swatch 24]

Acrylic

Acrylic is another DuPont fiber, created in 1950. The best substitute for wool, acrylic has superior thermal retention (retains heat), with good loft and bulk. The fiber is lightweight, and it can also feel as soft or softer than wool. Always in staple form (closer to wool), the fiber is used extensively for carpets, but it also appears in blankets, sweaters, mittens, and scarves, in blends with wool or alone. Acrylic offers warmth without weight and is washable, less expensive than wool, and hypoallergenic. For these reasons, the fiber competes really well with wool. Sunbrella® is a significant trade name for an acrylic that is used for outdoor furniture due to its superior outdoor wear and colorfastness to sunlight.

[Reference Swatch 25]

Modacrylic

Modacrylic is simply a modified acrylic. One of the significant modifications is the addition of a flame retardant chemical, which makes the fiber a little more wool-like. End uses for this fiber are fake furs, wigs, stuffed animals, paint roller covers, carpets, and scatter rugs. Modacrylic is also used for draperies and casements for public buildings due to its flame retardant status.

Polyester

Made of petroleum and first created in 1953, polyester is the most used manufactured fiber, garnering more than one-third of the fiber marketplace. An average weight fiber, polyester has the biggest variety of end uses, unlimited in its potential applications. Polyester is the great chameleon of fibers, able to successfully imitate silk, wool, and cotton, and it does so at a lower cost. There are more trade names, fiber variants, and new product developments in polyester than in all other fibers. These innovations have given the fiber new life, new possibilities, and a new-found respect. Fiber producers have engineered polyester to wick wonderfully and have a very pleasant hand. Consider well known fabric names such as Ultrasuede®, Polartec®, X-Static®, or Coolmax®.

Polyester has many outstanding qualities, with resilience (wrinkle recovery), dimensional stability (shrink resistance), and elasticity (recovery from stretching) being the most outstanding. This is the easy-care, wash-and-wear fiber. As the most commonly blended fiber, there are dozens of polyester and polyester blend swatches in this text; take the time to survey them and become familiar with the variations.

In hospitals, FR Polyester (permanently flame retardant) is quite desirable for privacy curtains and the like, due to the slow combustion, resistance to color loss by bleach, hardwearing durability, and low energy wash temperatures. It is hypoallergenic and recyclable.

Because of the great number of recycling programs available for polyester, it can be considered among the world's most eco-friendly fibers. Patagonia's Common Threads recycling programs have made a huge impact; EcoSpun®, Repreve®, Invista™, and Eco-fi® are just a few of the many trade names that reflect this pattern of ecologically sound decisions.

Polyester, and other synthetic fibers, are generally considered not sustainable, as we are using up finite petroleum reserves. We will not be able to reproduce or regrow these. Once depleted they are gone for good. Although only 10 percent is recycled today, all polyester can be recycled. By doing this, polyester can actually be a "greener" choice than many natural fibers, as there is little cost to the environment in recycling polyester. The use of clipping waste (loom and garment waste), along with recycled garments and a closed loop system, yields very usable recycled polyester, with little cost and no loss in quality. Recycled synthetics still use energy but with very little water and waste into the environment. All virgin polyester can be recycled using this method and can therefore be considered quite sustainable. These alternatives must be incorporated into design strategies as today's consumer will choose recycled over virgin alternatives.

[Reference Swatch 26]

Microfiber

Microfiber is sometimes made of polyester, but microfiber is a description of the size of the fiber (very thin, much finer than a human hair) and is not a generic fiber itself. Microfiber polyester (or rayon, nylon, acrylic, or modal) would be an appropriate descriptive term, but microfiber alone does not mean much. The properties of microfiber will be addressed in more depth in the next chapter.

[Reference Swatch 19]

Olefin

Also known as polypropylene and Tyvek®, olefin (developed in 1961) is the lightest-weight fiber and second in strength after aramid (below). It is also relatively inexpensive. For these reasons, the fiber is used for a Fed-Ex envelope, making them virtually impossible to tear and remarkably lightweight. This makes olefin a perfect choice, since the consumer pays for weight in shipping and the fiber is strong enough to survive the trip. You might recognize this fiber in wristbands placed on you at festivals. They are easy to cut off but very difficult to break or tear.

Olefin is also extremely hydrophobic and lighter than water. For this reason, it floats on water, and so it is the best fiber choice for all boating supplies, such as life jackets, ropes, and other boating paraphernalia. Olefin has a significant presence in upholstery, carpets, and outdoor furniture, again due to its hydrophobic nature. Hydrophobic means that it is stain resistant (to water-based stains).

Finally, olefin is often used as a house wrap to protect buildings under construction by providing a water barrier between the outer layers of a house and the frame and insulation. Water vapor can escape but liquid water cannot get in.

[Reference Swatch 27]

Aramid

Aramid is more commonly known by its trade names Kevlar® and Nomex®. Created in 1974, this fiber has few fashion or interior applications; it is primarily used in law enforcement and fire protective gear. Produced as bulletproof vests and flame retardant garments for race car drivers and pilots, aramid is the strongest fiber and only chars at a very high temperature. The fiber is relatively lightweight, which means that it does not add weight to a bulletproof vest, yet the fibers must be layered to be effective (to stop a bullet). Aramid is so strong that it is also used for reinforcing bridges and motorcycle helmets.

[Reference Swatch 28]

Spandex

Known as elastane in Europe, spandex is an elastomeric fiber, which means that it can stretch 100 percent and recover. This means that 5 inches will stretch to 10 inches. Spandex is never used alone; it is always used in conjunction with other fibers. As little as 2 percent spandex will add stretch, movement, and liveliness to any fabric. In interiors, a small amount of stretch fiber added to an upholstery fabric could make pulling the fabric over a frame significantly easier.

As its purpose is stretch, the fiber is always used in filament form. Lycra® is the most common trade name, produced by DuPont. Today, spandex has found its place in nearly every market.

Rubber

Rubber is the earliest elastomeric fiber. Spandex has three times the flex life of rubber and has superior longevity. However, rubber is still used in elastic waistbands and, most notably, scuba gear.

Table 4.1 shows a list of properties that can be associated with synthetic fibers in general. This is followed by a more specific list of properties attributed to each individual fiber, and finally the significance or definition of each property as it relates to the consumer (Tables 4.2 and 4.3).

The Burn Test

To differentiate one fiber from another, a burn test is often performed. Table 4.4 on page 38 will help you to begin to narrow down the identification of the fibers by their burn characteristics. For example, during a burn test, you discover that the fiber burns quickly with a bright orange glow. Further, that it continues to burn and smolder after it has been removed from the flame. In addition, the residue is ashy and the smoke smells like burning paper. With this chart, you will know that you have a cellulose fiber in your hand. All you have to do now is to narrow down and choose which cellulose: cotton, flax, ramie, or rayon.

Table 4.5 on page 39 will help you to identify the individual fiber. Observe the odor, the color of the smoke, and the residue to assist in identifying exactly which fiber you have just burned. Be careful!

Table 4.1 General Properties of Synthetic Fibers

Synthetics	General Advantages	General Disadvantages
Most synthetic fibers	Abrasion resistance Chemical reactivity: Excellent resistance Dimensional stability Drape Elasticity Environmental effects: Excellent resistance Hydrophobic* Resiliency Tenacity Thermal retention Thermoplastic Weight: Light to medium	Electrical retention Heat sensitivity Hydrophobic* Oleophilic *(oilophilic)*

*Hydrophobic is listed as an advantage because these fibers dry quickly and do not shrink or wrinkle due to their hydrophobic nature. Hydrophobic is also listed as a disadvantage because stains are not easily removed and perspiration is not absorbed.

Table 4.2 Properties Specific to Each Synthetic Fiber

Synthetics	Specific Advantages	General Disadvantages
POLYESTER Coolmax® Dacron® Eco-fi® EcoSpun® Gore-Tex® Invista® Polartec® Trevira® Ultrasuede® Many more	Abrasion resistance Chemical reactivity: Excellent resistance Dimensional stability Environmental effects: Biological resistance Resiliency Hydrophobic Tenacity Thermal retention Thermoplastic	Electrical retention Oleophilic *(oilophilic)* Pilling
ACRYLIC Acrilan® Orlon® Wear-Dated® Weather Bloc®	Dimensional stability Environmental effects: Resistance to moths Hydrophobic Lightweight Tenacity: Stronger than wool Thermal retention Thermoplasticity	Electrical retention Oleophilic *(oilophilic)*
MODACRYLIC SEF Verel®	As above for acrylic Flame retardant	As above for acrylic
SPANDEX Lycra® Cleerspan® Glospan®	Elasticity/Elastomericity 500%–600% elongation 99.9% recovery Filament only Lightweight	Abrasion resistance Bleach sensitivity Heat sensitivity Tenacity: Weakest fiber
ARAMID Nomex® Kevlar®	Abrasion resistance Chemical resistance Flameproof Tenacity: Strongest fiber	Pilling

(Table continued on next page)

Table 4.2 Properties Specific to Each Synthetic Fiber *(continued)*

Synthetics	Specific Advantages	General Disadvantages
OLEFIN Herculon® Marvess Polyethylene Polypropylene Propex® Tyvek®	Biological effects Chemical resistance Hydrophobic: 0% moisture regain Lightest weight Tenacity Wicking	Hydrophobic Oleophilic (*oilophilic*) Pilling
NYLON Polyamide Anso® Cordura® Gore-Tex® Supplex® Tactel® Zeftron®	Dimensional stability Hydrophobic Lightweight Resiliency Thermoplastic Tenacity Transparent Wicking	Hydrophobic Environmental resistance: Poor sunlight resistance Pilling

Table 4.3 Significance of Properties Common to All Synthetic Fibers

Properties	Importance to Consumer
Hydrophobic (low moisture absorbency)	Dries quickly; resists waterborne stains; lacks comfort in humid weather, as perspiration is not absorbed; increases the possibility of static. Water does not cause shrinkage or wrinkling; difficult to dye. Although generally stain resistant, these fibers are also difficult to clean.
Oleophilic *(oilophilic)*	Oil and grease absorbed into the fiber must be removed by dry cleaning solvents.
Electrical retainer	Static; shocks in cold, dry weather. Lint and dust are attracted and cling to surfaces.
Heat sensitive	If iron is too hot, fabric will melt.
Thermoplastic	Pleats, creases, and embossing can be heat set in fabric; fabric can be stabilized by heat setting; yarns can be textured for high bulk in carpets.
Chemical: Resistant to most chemicals	Wash-and-wear, easy care.
Environmental: Resistant to moths and fungi	Storage and longevity is not a problem.
Excellent abrasion resistance	Good appearance is retained longer because holes and worn places do not appear as quickly.
Excellent strength	Strong fibers resist breaking when stress is applied.
Excellent resilience	Wash-and-wear, easy care; less wrinkling during use.
Environmental: Excellent sunlight resistance	Not damaged by the sun. Good for window treatments, outdoor furniture.
Flame resistant	Varies from poor to excellent (check individual fibers); shrinks away from flame, melts, and drips into a plastic bead. Burns slow and self-extinguishes.
Weight	Varies as a group but tends to be lightweight.
Pilling	Occurs in staple-length fibers; fuzz balls.

Table 4.4 Burn Categories of Fibers

Flame Category	Burn Characteristics	Fiber Base
Inflammable	Burns quickly, smells like burning paper, leaves an ashy residue	Cellulose
Nonflammable	Does not burn	Mineral, aramid
Flame retardant	Burns slowly, sputters, self-extinguishes, leaves a crushable bead	Protein, modacrylic
Flame resistant	Melts and shrinks away from flame, self-extinguishes, leaves a hard plastic bead	Synthetics: Thermoplastics

Table 4.5 Burn Characteristics of Fibers

Chemical Composition	Fiber	In Flame	Odor	Residue
Cellulose	Cotton Flax (linen) Ramie Hemp	Burns quickly, with a luminous orange glow, smolders after flame	Burning paper	Light, feathery, gray ash; linen has a whiter ash
Manufactured cellulose	Rayon Bamboo (rayon) Cuprammonium Lyocell Modal®	Burns with a luminous orange glow, smolders after flame	Burning paper	Small, gray, feathery ash
	Acetate	Burns and melts rapidly, may drip, also has ash	Vinegar, acidic	Ashy, with hard, shiny black bead
	Sorona® (corn) Cocona® (coconut)	Burns, drips, sputters, melts, and produces black smoke	Sweet, chemical	Hard, shiny black bead
Protein	Wools (mohair, merino, cashmere, camel, alpaca, llama, and so on)	Burns slowly, sputters, self-extinguishes, scorches	Burning hair	Gray-black bead, easily crushed to black powder
	Silk	Burns slowly, sputters, self-extinguishes, scorches	Burning hair or feathers	Gray-black bead, easily crushed to black powder
Manufactured protein	Soy	Burns slowly, produces white smoke	Burning hair, vinegar	Textured, plastic, black beads
Chemical	Nylon	Melts before igniting, drips, sputters, self-extinguishes, produces white smoke	Celery	Hard gray-brown or yellow bead
	Polyester	Melts before igniting, burns with a heavy black smoke	Sweet, chemical, oily	Hard black round mass
	Acrylic	Melts and burns, may produce black smoke	Sour or similar to burning meat or broiled fish	Hard, irregular, brittle black ash & bead
	Modacrylic	Burns very slowly, self-extinguishes	Sharp; gunpowder	Black ash, may be a black bead
	Aramid	Does not burn	N/A	N/A
Mineral	Glass and metallic	Does not burn	N/A	N/A

Activity 4.1: Fiber Burn Test

Name: ______________________________ Date: ____________

Objective: To identify fibers by their chemical composition by observing their burn characteristics. You will quickly see the differences between fibers as you perform these tests.

Materials Required:

Fabric swatches (see the following fiber burn test chart), tweezers, an unscented tea candle, aluminum foil or a commercial aluminum pie plate (perhaps both), scissors, and the fiber burn test chart. Note: Two coins or keys may be used in place of tweezers, but tweezers are best.

Perform this test indoors in a safe, well-ventilated area, where you are unlikely to set off smoke alarms.

Procedure:

1. Light the candle. Place it in the middle of the aluminum pie plate or a piece of foil.
2. Cut a ⅛-inch strip from each of the swatches referenced in the fiber burn test chart. (Note: It is more accurate to burn only a cluster of horizontal or vertical yarns of the fabric, because a blend is possible.) Notice the differences between fibers in the burn test.
3. Holding the ⅛-inch strip with the tweezers, place the end of the fabric in the flame. Make sure that it actually catches fire (information cannot be obtained if the fabric strip is allowed to shrink and melt away from the flame).
4. Observe the nature of the burn (does it ignite quickly?) and the color of the smoke. Also observe what happens when you remove the fabric from the flame: Does the fire go out immediately? Or does it continue to burn? Record all of these observations on the form provided.
5. Blow out the fire if necessary and quickly sniff the smoke for the odor of the burned fibers (make sure that the fire is out before you do this; there are some rather sensitive hairs [fibers] in your nostrils).
6. Once the residue has cooled, feel it to determine if it is a hard plastic bead, a crushable bead, or simply ash. Be careful! Do not touch the residue until it has cooled. Some of these fibers can create a nasty burn!
7. Record your results in the fiber burn test chart.
8. Attach the burned residue in the last column.
9. The fiber content of the last two swatches are not listed in your book. Perform the burn test and identify the fiber content of them.

Activity 4.1 Fiber Burn Test *(continued)*

Name: ______________________________ Date: __________

Comparison of Fibers by Burn Test

Swatch #	Fiber Content	Burning and/or Melting	Color of Smoke	Odor	Type of Residue	Actual Residue
5	Cotton					
7	Flax					
8	Hemp					
11	Wool					
13	Silk					
65	Rayon					
18	Bamboo					
20	Acetate					
21	Corn					
22	Glass					
24	Nylon					
25	Acrylic					
26	Polyester					
27	Olefin					
28	Aramid					
97						
99						

Further testing includes solubility testing: Acetate will dissolve in acetone solution (nail polish remover); protein fibers (wool and silk) will dissolve in bleach.

Activity 4.2 Fiber Application Assignment

Name: ______________________________ Date: __________

Objective: To justify the use of specific fibers for a given end product, based on evaluation of their properties.

Procedure

1. Research current ads in favorite websites to find three different textile based products. The ads must include an image and the specific fiber content for each.
2. The three ads must be for three different interior design products, made with different fibers.
3. The three fibers should be across chemical bases. For example, if one is a cellulose, then another might be a protein, and the third a synthetic or perhaps a regenerated cellulose.
4. To reduce complications, avoid blends. Find ads with single-fiber fabrics.
5. **Property names:** Using formal properties discussed in Chapter 1, identify the three most positive properties for the fiber featured in each of the ads. Select properties that reflect the best qualities of the fiber to validate consumer use.
6. **Definitions of properties:** Define each of the three properties identified for each ad, in your own words. Finally, explain why the property is relevant to the product's end use.
7. Presentation should include the following:
    - A title page with the title of the project, your name, your instructor's name, the course number, the class day and time, and the date.
    - Each ad should be on a separate 8½ × 11-inch page. The ads should include the text from the ad describing the product and the fiber content. Highlight the text that describes the fiber content.
    - Follow the ad with your typed, double-spaced content, which should appear on the same page as the ad.
    - Staple the pages together; no folder required.

Guidelines:

- Choose properties relevant to the consumer. (For example, the most relevant property of a towel is absorbency [hydrophilic]).
- Avoid properties that are average. Sell this product with the best possible properties!
- Write about fiber, not fabric!
- Avoid trade names; this is a generic fiber assignment.
- Use formal properties: Cheap and washable are not properties.
- Each of the three ads must feature a different fiber, and each ad must be for a different product. (Example: For a bathroom, towels, window treatments, and a rug.)

Sample Project

Name of product: Bath rug
Specific fiber content: Olefin

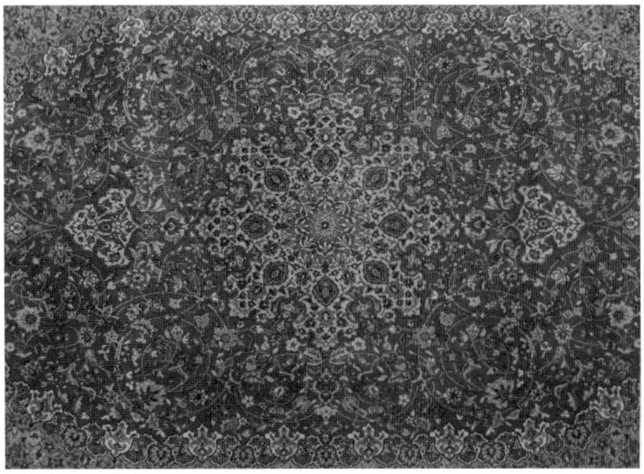

1. Hydrophobic, which means that it is not absorbent. This is a desirable property for a bath rug because the rug does not hold on to water created by a bath or shower.
2. Wicks well, which means that it will transfer any water to the surface for immediate evaporation. If water is retained in a rug, mildew or rot may occur. Because olefin wicks well, it is not susceptible to mildew.
3. Dimensional stability, which means that this fiber will not shrink or lose its shape. This is important because bath rugs are often laundered. The consumer will get great wear out of this rug because it will retain its shape and size through launderings.

CHAPTER FIVE

Yarn Classification

Reference Swatches

29. Single Multifilament | 30. Single Spun | 31. Two-ply Spun

32. Bouclé | 33. Chenille | 34. Chenille Damask | 35. Tweed

Yarn Classification

A yarn is a group of fibers twisted together to form a continuous strand. Yarns can be filament, spun, or complex.

After fiber analysis, yarn construction is the next step in textile production. A thorough understanding of yarn characteristics is critical in predicting the textile's performance and will help make you a better designer. For example, fabrics made from filament yarns offer superior drape and luster. They also tend to cause more static cling. Spun yarns provide better absorbency but will also pill. Decisions, and trade-offs, must be carefully considered (Figure 5.1).

Filament Yarns

Filament yarns are made from very long, filament, manufactured fibers and silk. A group of filament fibers are laid together to make a yarn. *Filament fibers make filament yarns.* These yarns make smooth, lightweight fabrics such as drapery sheers and satin sheets.

There are two types of filament yarns.

- Monofilament yarn is a yarn made of a single fiber. It is stiff, not flexible, and is rarely used in fabrics; notable exceptions are metallic fibers and spandex.
- Multifilament yarn is made of many fibers. In carpeting, there could be as many as 100 fibers in a yarn. Fabrics made of multifilament yarns are flexible, drapable, lustrous, and smooth, with a cool hand. Dental floss is a good example of this yarn type. Multifilament yarns can be flat (smooth and lustrous) or textured (crimped). Textured filaments offer the fabric improved drape, resilience, and elasticity. They also provide a significant contribution for carpeting yarns as they can be bulked up and provide great covering ability. A textured filament also improves warmth, reduces pilling, and makes the synthetic fiber look and feel more natural.

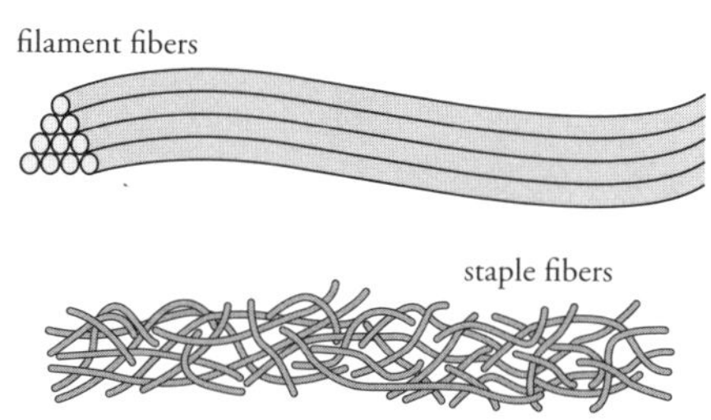

Figure 5.1 Filament and spun yarns:
(a) Filament fibers become filament yarns.
(b) Staple fibers become spun yarns.

Spun Yarns

Short, staple fibers must be twisted or spun to make yarns. Most natural fibers are staple. Manufactured fibers are extruded in filament lengths but may be cut to staple and spun as well. *Staple fibers make spun yarns.* Fabrics made from spun yarns are soft, fuzzy, warm, more elastic, and resilient with a warm hand. See Table 5.1.

For a direct comparison of filament versus spun yarns, see Swatches 29 and 30. Reference Figure 5.2a and b for a visual comparison of spun and filament.

Processing Spun Yarns

Staple fibers are first **carded**, which means brushing, disentangling, and laying the fibers straight. Once carded, the fibers are twisted together to become spun yarns. As an optional second step, **combing** is possible. This creates a finer fabric. Since the shorter fibers are removed, and the fibers are laid more parallel, a fine smooth yarn is produced. Think of brushing versus combing your hair. Combing gives you a finer result. Combed yarns are made of cottons and provide smooth, fine grade sheets, such as percale. See also Swatch 40, Organdy, for an example of a **combed cotton**. Most cottons are carded only, as in canvases, muslin, and T-shirts. See Swatch 39, gauze.

Woolen yarns are made of wools and are carded only. They are warmer and fuzzier, with more bulk, and are best used for warm blankets and carpets but also soft, inviting upholstery fabrics. See Swatch 60, Houndstooth, for an example of a woolen yarn and compare it to Swatch 77, which is a much finer **worsted** yarn. Worsteds provide a cleaner, less fuzzy, smoother result. Worsted wool means that the yarn has been carded and combed.

Table 5.1 Properties of Spun and Filament Yarns

Spun Yarns	Filament Yarns
Thermal retainer	Thermal conductor
Warmer, softer hand	Cool hand
Fuzzy, bulky, lofty	Fine, smooth, less bulk
Elasticity	Poor elasticity
Hydrophilic	Less absorbent
Resilient	Poor resilience (wrinkles)
Less strong	Stronger
Shrinks	Dimensional stability
Tends to pill and lint	Less pilling and lint
Soils easily	Sheds soil
Does not snag easily	Snags Less
Better cover	Cover
Poor drape	Excellent drape
Poor luster	Excellent luster
Electrical conductor	Electrical retainer
Size expressed in yarn count	Size expressed in denier

Yarn Types

Once spun and filament yarns are created, they are sometimes plied. Both yarn types may be left as singles or plied. Single-ply yarns are the result of the first twisting operation; a group of fibers are twisted into a single yarn. Two- or more-ply yarns consist of two or more single yarns twisted together to form a new yarn.

Plying adds strength. The higher the ply, the stronger the yarn, but a high ply number does not necessarily mean that the yarn is thicker. *Two-ply* means two separate yarns are twisted together into a single yarn, and *three-ply* means three separate yarns are twisted together into a single yarn (see Figure 5.2e and f). Reference Swatch 62.

If plied yarns are twisted together, they become a cord, substantially thicker than a plied yarn, and often used for decorative purposes. See Figure 5.3.

Yarns may also be core spun. Core-spun yarn consists of fibers twisted around a central core. Traditionally, core-spun yarns have consisted of cotton wrapped around a poly core, but today many stretch fibers are incorporated into a fabric in this way. This method of creating yarn results in fabrics that have more stretch and are softer, with more strength and absorbency.

Novelty or Complex Yarns

Novelty or complex yarns are a combination or variation of spun and filament yarns. These yarns are decorative, adding interesting texture to the surface of a fabric. They are often not durable, the yarns are mostly used in the horizontal direction of fabrics, as they are not strong enough to be used in the vertical, which takes the stress and abrasion of the weaving process. Novelty yarn types include slub, bouclé, nub, chenille, flake, spiral, ratiné, loop, curl, knot, spot, and metallic. These are used in decorative pillows or throws on the sofa, or the bottom of the bed. They are textured and soft and usually quite seductive, although not durable. See Figure 5.2c and d for an example of chenille and bouclé.

[Reference Swatches 32–35]

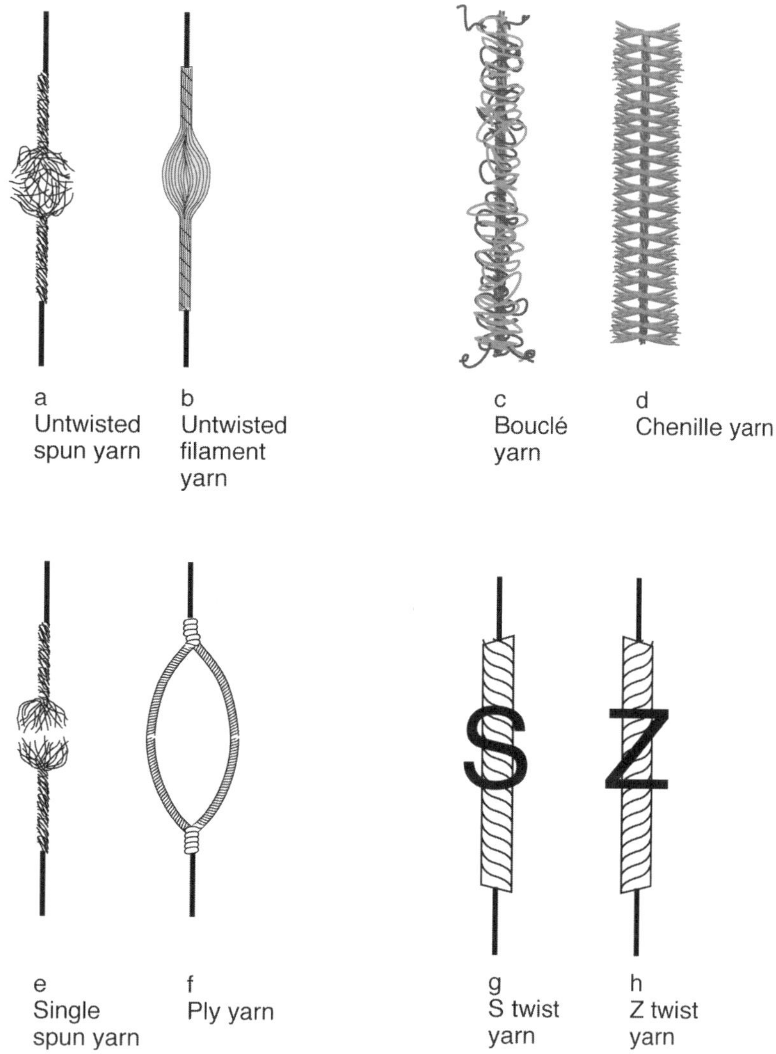

Figure 5.2 Yarn variations. (a) untwisted spun yarn (b) untwisted filament yarn (c) bouclé yarn (d) chenille yarn (e) single-spun yarn (f) two-ply yarn (g) S-twist yarn (h) Z-twist yarn

Yarn Twist

The strength of a spun yarn is determined by three criteria. The first is the fiber content; stronger fiber means stronger yarn. Second, the amount of crimp in the fiber determines the cohesiveness of the yarn. Cohesiveness assures that the fibers will cling together to form a better, stronger yarn. Finally, a measure of the number of twists or turns per inch (tpi), can be used as a determinant for strength. Twist holds the fibers together but also adds strength to the yarn. Essentially, more twists (higher tpi) result in a stronger yarn. Table 5.2 illustrates the properties affected by an analysis of twist.

A low-twist, soft-spun yarn, with 2 to 12 tpi, has less twist and is less compact (bulkier). It is also weaker, less elastic, and has a softer, warmer, and fuzzier hand. Low-twist yarns are slower drying, duller, and less durable, but they are more absorbent. Woolen fabrics are made with low-twist yarns. See Swatch 35.

A high-twist, or hard-spun, yarn has 20 to 30 tpi. It will have more twist and be more compact. High-twist yarns are stronger, harder, smoother, more elastic, and more durable. They are also less insulating. Yarns with more twist have a cooler hand, and are less absorbent, which means they dry more quickly. Filament voile has hard-twisted yarns. See Swatch 37. Worsted fabrics are made of high-twist yarns.

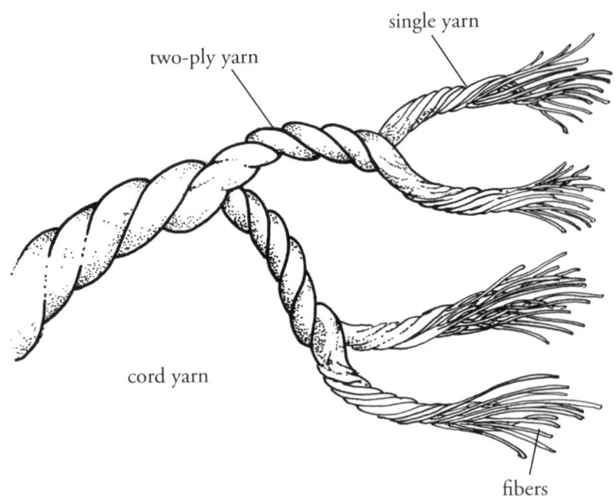

Figure 5.3 Yarn structure

Over-twist, or crepe twist, yarn has more than 40 twists per inch. This kind of yarn is not as strong or lustrous as lower twists but has more elasticity, more drape, and is more resilient. Chiffon, georgette, and crepe-back satin are fabrics made of over-twisted yarns. A less expensive alternative to over-twisting yarns is a textured filament. Thermoplastic (synthetic) fibers can be given a tight crimp, which increases elasticity, bulk, drape, and texture.

Yarns can also be evaluated by the direction of the twist, illustrated in Figures 5.2g and h. Texture is often added to a fabric by including both S & Z twists. Georgette is one such example. Reference Swatch 44.

Table 5.2 Properties Affected by Analysis of Twist

Low Twist (2-12 tpi)	High Twist (20-30 tpi)	Over-Twist (over 40 tpi)
Soft-spun yarn (i.e., woolen)	Hard-spun yarn (i.e., worsted)	Crepe-twist yarn (i.e., chiffon)
Warm hand	Cool hand	Rough hand
Soft & fuzzy	Fine & smooth	Messy appearance
Thermal retainer	Less insulating	Not insulating
Hydrophilic	Less absorbent	Not absorbent
Slower drying	Quick drying	Fast drying
Bulk	Less bulk	Very fine
Luster (if filament)	Less luster	No luster
Poor elasticity	Elasticity	Excellent elasticity
Poor resilience	Improved resilience	Excellent resilience
Low strength	High strength	Poor strength
Less drape	Improved drape	Excellent drape

The construction and direction of twists of yarns can impact the performance and appearance of the final product. When we change the yarn type, we can make a linen with reduced wrinkling, or even retard shrinkage in a cotton fabric. Although polyester is known to pill, spun polyester pills much more readily than filament polyester. Spun polyester is also warmer and softer, with more bulk and elasticity. Filament polyester drapes better and has a cooler, slicker hand, but it snags more easily.

Yarn Performance

Different yarn structures can have a dramatic impact on performance and appearance. Properties don't end with fibers, they extend to yarns and even fabrics. Filaments are the most lustrous, and spun yarns are soft and warm, while novelty yarns have the most surface interest. Yarns can also influence the propensity of a knit fabric to run. Spun yarns resist running; a textured filament will resist running almost as much, while slippery filament yarns run easily. With all this in mind, your choice for a particular fabric must extend beyond fiber and fabric to yarn construction.

Sometimes the properties of yarn construction are at odds with the fiber or even the fabric, but in order to predict precisely the performance of a given fabric, all elements of the textile must be considered. Each of the fabrics in this chapter offer an example of a basic yarn type. As we evaluate these yarn constructions, attempt to disregard the fiber content and fabric construction. For the purpose of this chapter, we will be looking at the yarn construction only.

Yarn Sizing

Another criterion used to evaluate fabric for a specific use is yarn size. Today, more than ever before, we are making purchases online, even for larger purchases such as upholstered furniture. As a designer, it is helpful to understand the language to assist in making good decisions. Understanding the size of the yarn that the fabric is made of gives the buyer information about the hand, weight, and fineness of the fabric. Three common systems are currently used to express the size of the yarn: yarn count, denier, and tex. Ultimately, the finer the yarn, the finer and smoother the fabric.

Yarn Count System

The yarn count system is an indirect system used to express the size of spun yarns. This system is used to identify the size of a natural yarn. An example is *10/2*, which means "size 10, two-ply." A bigger number means a smaller yarn. In this case, two yarns, each size 10, are twisted together to form a ply yarn. The ultimate size of the yarn is equivalent to a 5/1. See Swatch 2 for an example of yarn measured by yarn count. The yarn size of wool or wool blends is expressed in the reverse, *2/10*, with the same meaning attached.

The Denier System

Denier is a system of yarn sizing that represents the weight of 9,000 meters of filament yarn (or fiber) expressed in grams. This is a direct system that applies to manufactured fibers only. Direct means that the bigger the number, the bigger (thicker) the yarn. A semitransparent café curtain might have yarns made of 120 denier, panty hose might be 10 denier, while carpeting can be 2,500 or more. The finer the denier, the softer the hand. As filament yarns are rarely plied, denier is represented as a single number.

Microdenier/Microfiber

Microfiber (microdenier) is not a fiber; it relates to the size of fiber. The definition of a microfiber is any fiber that has been extruded thinner than 1 denier in size (between 0–1 denier). Microfiber can be made of any man-made fiber, acrylic, polyester, nylon, rayon, or modal. Microfibers provide superior wicking, drape, and a very silky hand. Nanofibers are much finer than microfibers. Swatch 3 is a yarn represented in terms of denier.

The Tex System

The tex system (often represented as Dtex) is an international system based on metric weight. Like denier, it is a direct system; however, the two systems (denier and tex) are not equal to one another. All categories of yarn can be measured using the tex system and assigned an appropriate number for the purpose of clear, even communication. In the tex system, the bigger the number, the bigger the yarn. Swatch 3 is also represented in tex.

Use Table 5.3, Yarn Study Guide.

[Reference Swatches 29–35]

New Innovations in Yarns

Much of the high performance marketplace begins in the yarn stage. Technologies such as Celliant and Thermocules are both related to body temperature regulation, and both are engineered at the yarn level.

Sewing Thread

Sewing thread is a subset of yarns, used to connect product parts together. For upholstery use, the thread is plied (2 or 3 ply). Finishes are also of crucial importance. Glazing is a finish that assists in making the sewing thread stronger.

Table 5.3 Yarn Study Guide

Categories of Yarns	
Filament yarns	All man-made fibers and silk. Filament fibers make filament yarns.
Spun yarns	All natural fibers except silk. Silk noil and man-made fibers cut to staple length are also spun. Staple fibers make spun yarns.
Novelty yarns	Slub, bouclé, nub, chenille, flake, spiral, ratiné, loop, curl, knot, spot, metallic
Properties of yarns	
Filament yarns (fabrics are silklike)	Smooth, lustrous, no pilling or linting, shed soil, cool, slick hand, less cover, little loft or bulk, stronger, better drape
Spun yarns (fabrics are cotton, wool, or linen-like)	Dull, fuzzy, subject to pilling, lint, soil easily, warm hand, do not snag easily, more cover (opaque), more loft and bulk, resilient, elastic, less strong
Novelty yarns	Purely decorative effects—not durable. Mostly used in the horizontal direction of fabrics as they are not strong enough to be used in the vertical direction.
Yarn Types	
Filament yarns	Subdivided into two types: • Monofilament: a yarn made of a single fiber. Stiff, not flexible, rarely used in apparel fabrics. (Notable exception is spandex and metallics.) • Multifilament: a yarn made of many fibers. Flexible, drapable, smooth hand, lustrous.
Spun and filament yarns	Further subdivided into: • Singles: the result of the first twisting operation. Fibers twisted into yarn. • Ply: two or more yarns twisted together to create one yarn. Two-ply means two separate yarns twisted together into a single yarn. Three-ply means three separate yarns twisted together into a single yarn.
Amount of Yarn Twist (TPI) TPI = twists or turns per inch **Low: 2–12 tpi (soft twist); High: 20–30 tpi (hard twist); Over-twist: >40 tpi (crepe twist)**	
Soft-spun yarns	Less twist, less compact, weaker, less elastic, softer, warmer, fuzzier, slower drying, duller, less durable, more absorbent. (Example: woolens)
Hard-spun yarns	More twist, more compact, stronger, more elastic, harder, smoother, more lustrous, less insulating, cooler hand, less absorbent, quicker drying, more durable. (Example: worsteds)
Crepe-twist yarns	Less strength, less luster, grainy hand, more elasticity, more drape, more resilience. (Example: georgette)
Yarn Sizing	
Yarn count system	The size of spun yarns, expressed in numbers, called yarn count. Bigger # = smaller yarn. 10/2 means size 10, two-ply. 20/2 is finer than 10/2
Denier	The weight of 9,000 meters of filament yarn (or fiber) expressed in grams. Bigger # = bigger yarn.
Tex system	International system of measuring based on the metric system of weight. All categories of yarns can be incorporated into this system and can be assigned an appropriate number for purposes of clear communication. Direct system. Bigger # = bigger yarn. (Note: Although both tex and denier are direct systems, they are not equal to one another.)

Activity 5.1 Swatch Evaluation: Yarns

Name: ____ Date: ____________

Evaluate swatches by their yarn types. Read the instructions and hints carefully!

Procedure:

Using the horizontal yarn in each swatch, determine if the yarns listed below are spun or filament. Place your answer in Column D.

If filament, address whether the yarn is monofilament, multifilament, or a textured filament. Place your answer in Column E. For spun yarns, you need not fill in this box.

Identify the yarn type. Is the yarn constructed as a single or plied yarn?

Place your answer in Column F. (Hint: Do not confuse a single multifilament—by definition, one group of many yarns—with a two-, three-, or four-ply yarn.)

If the yarn is plied, how many ply? Place your answer in Column G. Only fill in this box if the yarn is, in fact, a plied yarn.

Swatch Evaluation: Yarn

A. Swatch	B. Fabric Name	C. Fiber Content	D. Spun or Filament	E. Mono-, Multi-, or Textured Filament	F. Yarn Type *Single or Plied?*	G. # of Plies
6	Jacquard	Cotton				
13	Novelty rib	Silk				
14	Houndstooth	Silk				
20	Satin	Acetate				
23	Basket weave	*Metallic (only)				
25	Twill weave	Flannel				
46	Monk's cloth	Polyester				
39	Gauze	Cotton				
41	Organza	Nylon				
63	Satin	Polyester				
75	Crepe	Rayon				
79	Jacquard	Rayon/ Poly (orange horizontal)				
140	Glazed Chintz	Cotton				
143	Georgette	Polyester				

Note: These elements are blank on the corresponding swatch boards. Once completed and approved by your instructor, add the missing information.

- For swatch #23—evaluate the metallic yarn only
- For swatch #79—evaluate the orange horizontal yarn only

CHAPTER SIX

Plain Weaves

Reference Swatches

36. Voile | 37. Filament Voile | 38. Crushed Filament Voile
39. Gauze | 40. Organdy | 41. Organza | 42. Gingham
43. Chambray | 44. Shantung | 45. Crash/Butcher Linen

Fiber and Fabric

The difference between fiber and fabric is one of the most fundamental concepts in textiles and generally one of the most misunderstood. Fibers are the basic building materials. They are small and hairlike, the smallest visible components of a textile. Fibers are made into yarns, and yarns are made into fabrics. Fabrics are the result of weaving, knitting, or other methods of construction, and they are subsequently used to upholster a chair or drape a window. Weaving is the oldest method of making a fabric.

Identifying the fiber content is significant to the performance of a product, but it does not tell the whole story. Think of describing a house by calling it wood. While the house may be built primarily of wood, that description does not create an adequate representation of the entire structure. Quite simply, fiber is what the fabric is made of, and fabric is what the fiber is made into. Understanding the relationship between fiber and fabric is essential to every textile-related discipline from product development to interior design. Cotton is a very common fiber that can be constructed into many different fabrics. Yet not all cotton fabrics are alike in their performance capabilities. Contrast denim with a tablecloth damask or corduroy.

Silk and satin are often confused; they are not the same thing. Silk is a fiber, and satin is a fabric. Sometimes silk is made into satin, but more often than not, satin is made of rayon, acetate, nylon, or even polyester. And silk can be made into many other fabrics, such as brocade or crepe de chine.

Fabric stores often organize their inventory by fiber. If you ask an employee the name of a fabric, he or she might tell you it is cotton or silk, while showing you a jersey or broadcloth. Although this practice is common, it leaves too much room for expensive errors. The fiber content is not the fabric name. Think of all the cotton fabrics that you know. Are they interchangeable? If you decide to use cotton for a tablecloth, and you order cotton fabric, could you still use it if you received cotton corduroy or cotton jersey? No, and that is why it is not enough to identify a fabric by its fiber content. Knowing the fiber content is a good beginning, but for people working in the industry, a greater knowledge of fabric names and their performance criteria is required to successfully create a textile product for fashion, home, or industrial applications.

Care labels on a pillow, bedspread, garment, or other product are rather limited in scope. If a label says cotton, silk, polyester, or rayon, it is the fiber content that is listed and not the fabric. In fact, fabric names are not required on most labels. Although absent, fabric identification and the performance prediction associated with the fabric is critical to the success of the product.

Review Swatch 1 (cotton fibers) and Swatch 4 (the simplest cotton fabric) for a reminder of the difference between fiber and fabric.

Identifying Fabrics

Most people attempt to identify fabric by touch. With today's textile manufacturing technology, one can be completely fooled by this method. At best, fiber can be identified this way, although this is extremely unreliable, even for fibers. A closer look, and a more analytical approach is required to accurately identify and differentiate one fabric from another.

As we move through this section, we will be looking at many fabrics that have the same basic fabric structure but different names. In order to articulate one from another, we must first define the criteria for identifying fabrics so that recognition becomes possible.

Criteria for Fabric Identification

The names of fabrics can be tied to their characteristics, finishes, structure, origins, pattern, and even performance. Understanding this concept can be helpful in determining a fabric's identity. The following are naming conventions for fabrics.

- Fabrics can be named for their inherent structure. Herringbone is always a reversing twill; the structure relates to the fabric's method of construction and is created by a specific interlacing pattern. (Interlacings are the organization of horizontal and vertical—warp and weft—yarns as they cross over and under one another.)
- Fiber content is sometimes responsible for a fabric's name, such as China silk or handkerchief linen. Linen is a plain-weave fabric made of a bast fiber, usually flax, but not always. China silk is always a plain weave made of silk.
- Fabrics can be named for their origins, such as damask for Damascus, paisley for the city in Scotland, and gauze for Gaza.
- Fabrics are often distinguished by their finishes. Flannelette is generally a simple plain-weave cotton fabric. With a napped finish it becomes flannelette and performs differently. Without this finish, it would not be flannelette.
- Method of coloration can be a factor in identification. Gingham, chambray, and madras are all plain-weave cotton fabrics that are in the same weight category. The differences between them lie in the organization of their yarn-dyed colors. Madras uses several yarn-dyed colors in both directions, creating a plaid design. Chambray has white yarns in one direction and colored yarns in the other. It is often mistaken for denim, which has the same color organization but a different structure. Gingham uses alternating blocks of white and a color in both directions to form a checked pattern.
- Yarn construction can determine fabric identity. Voile and batiste are basically the same fabric, with differently twisted yarns. Voile has hard-twisted yarns; batiste is much softer. The placement of spun and filament yarns can determine fabric name as well. Bengaline is often identified by a spun weft and filament warp.
- Print can be a determining factor. Calico is a lightweight, plain-weave, cotton-like fabric with a small floral print. Although each of these criteria is important, it is ultimately the print that gives this fabric its identity and distinguishes it from many other fabrics of like hand and structure.
- Finally, weight is very important in fabric identification. Often, fabrics with different identities share the same fiber, yarn, and fabric construction. Poplin and broadcloth are basically the same fabric; their distinguishing feature is simply weight. Broadcloth is a lighter-weight version of poplin.

The last factor in this list, weight, is one of the most significant criteria of a fabric. Fabric is often bought and sold by weight, rather than yardage. Determining product use is due, in large part, to the fabric's weight. In fact, often the only difference between apparel fabric and fabric more appropriate to interiors is weight. Likewise the difference between fabrics for windows and for upholstery is often nothing more than weight.

Consider the case of denim. This is an extremely common fabric that everyone knows, but not all denim is the same. Denim is described by weight, often expressed as ounces per square yard (oz./sq. yd.). It can also be described as grams per square meter (gms./sq. m). Denim can range from a very thin, light weight of 4 to 5 ounces per square yard up to the aptly named bull denim, which can weigh as much as 22 ounces per square yard. Different weights of fabrics have different uses. Table 6.1 shows appropriate uses for fabrics of different weights.

Table 6.1 Basic Weight Categories

General Category	oz./sq. yd.	Appropriate End Uses
Extremely lightweight	0–1	Sheers, netting, meshes
Lightweight	1–6	Sheers, sheets, casements, laces, table linens, light curtains
Medium weight	7–12	Brocades, jacquards, drapes, lightweight upholstery, toweling, canvas, blankets, quilts
Medium–heavy weight	13–19	Heavy upholstery, light carpeting
Heavy weight	> 20	Carpeting

Anatomy of a Woven Fabric

Woven fabric is made with a series of interlacings: two sets of yarns at right angles to each other. The vertical yarns are called the warp. They are also called ends. The horizontal yarns are called weft, picks, or filling. The terms are used interchangeably; their use is dependent on the segment of the industry one is conversing with. There is usually (but not always) more warp than weft in a fabric. The warp direction of a fabric is placed in the vertical direction on a chair or sofa, as there is more strength in this direction of the fabric. The warp threads are set up on a loom (which is essentially a weaving machine), which creates tension, to make weaving easier. Refer to Figure 6.1 for an example of a simple, schematic loom. The weft or filling threads fill the warp, completing the fabric.

The finished edges of a fabric (that don't unravel) are called selvages. The selvages are densely woven to discourage fraying or unravelling. The selvages must be cut away for the fabric to properly drape. The grain line of a fabric relates to the warp direction, which is parallel to the selvage. The cut edge of a fabric is parallel to the weft yarns.

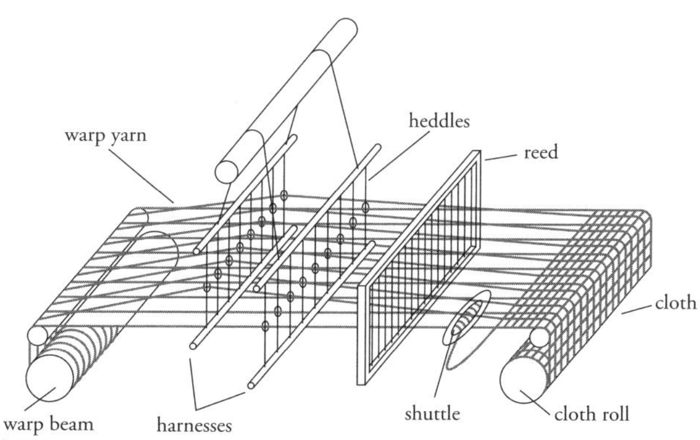

Figure 6.1 Loom

Fabrics have a front, called the technical face, and a back. In many instances, such as sheeting, it is difficult if not impossible to identify the front from the back. In others, it is easily apparent and critically important to identify for proper application.

Fabrics can be differentiated from one another by identifying one of the many patterns of interlacings, also called weave structures. This chapter analyzes the first and simplest structure, plain weave.

Qualitative Analysis of a Fabric

One of the determinants of quality in a fabric is **thread count**. Think about bedsheets for a moment. The price of sheets is determined by thread (yarn) count. The higher the count, the more you pay for sheets. Thread count is the number of threads in each and every square inch of the fabric, so an 800 thread count, for example, is quite impressive.

To determine thread count, line the pick glass (also called a linen tester) up along the warp and weft of a fabric and begin by counting all the vertical yarns in this window. This can be difficult to see and is aided by a pick needle, pin, or even a pen. You must count all of the threads in this inch, even the ones that are under others. Sometimes it is easier to pull yarns out of the bottom of the swatch, creating a fringe. Then count the fringe. Once you have the number of warp threads, count the horizontal yarns. Try counting Swatch 26, which has an easy count of 34 × 13. This is a total count of 47 yarns in this inch—not very impressive in sheets, but easy for you to count. Unlike sheets, counts are always expressed as both warp and weft and not the sum. The response would be 34 × 13, not 47.

Balance is the ratio of warp to weft. A balanced fabric has an approximately equal number of warp and weft threads. Even in a balanced fabric, there is usually more warp than weft. Swatch 26 is an unbalanced fabric, with twice the warp number. A balanced fabric has improved abrasion resistance and dimensional stability, but often decreased drape and resilience.

Additionally, you can see how easily this fabric can fall apart due to its low count. **Count** is the actual number of yarns in both warp and weft. A higher count will impact a fabric in many positive ways, improving drape, dimensional stability, and hand. Sometimes, as in a satin, for example, a high but unbalanced count would be desired for luster, drape, and resilience. When evaluating count, bigger is better.

We now have criteria with which to determine if a fabric selection is right for its intended use. We have considered fiber content and yarn construction. A detailed analysis of balance and count are also essential in determining a fabric's viability. Weave structure is as important as fiber content in predicting the final performance of the product. Properties attach to structure as much as to fiber.

In making the ultimate decision if a fabric will work for a particular purpose, consider the end use. How will this fabric be used? If the fabric is chosen for a drape, then sunlight resistance and drape are of the highest significance. Abrasion resistance and strength are of paramount importance for upholstery applications. A decorative fabric might be suitable for light duty, for example, a decorative chair that is hardly used. Moderate duty would include a formal sofa in a living room that is seldom used. A more durable fabric might be required for a favorite reading chair, while durability must be considered for the TV room. These considerations must be applied to all the elements of a textile, beyond fibers through yarns and fabrics.

When purchasing interior fabrics, the term **railroading** may be applied. This means that the fabric may be used horizontally instead of vertically for maximum efficiency. Essentially, the fabric can be used sideways. **Up the bolt** means that the fabric must be used vertically. Reference Figure 15.2a & 15.2b.

Plain Weaves

The first fabric structure for analysis is a plain weave, which is identified by a simple 1:1 ratio. The two sets of yarns interlace at right angles in an over-one, under-one construction, creating a checkerboard appearance, as in Figure 6.2. Because of the 1:1 ratio, the count of this fabric is well balanced. As it is balanced, the front and back of a plain weave are the same, unless one side has been printed or finished. Structurally, they are the same. This is the least expensive weave to produce and therefore the most common weave structure.

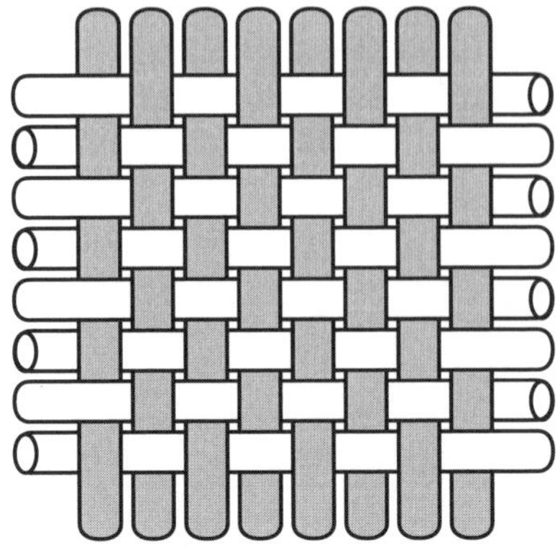

Figure 6.2 Plain-weave structure

Plain-weave fabrics share the following performance expectations:

Durability

- Excellent abrasion resistance (resists abrasion)
- High tensile strength (hard to break)
- Low tearing strength (easy to tear)
- Good dimensional stability
- Poor elasticity
- Snag resistance
- Light weight (depending on fiber, yarn size, and construction and count)

Care

- Poor resilience

Appearance

- Low luster
- Poor drape
- Good body

[Reference Swatches 36–45]

Use Figure 6.3 as a quick reference for identifying plain weaves.

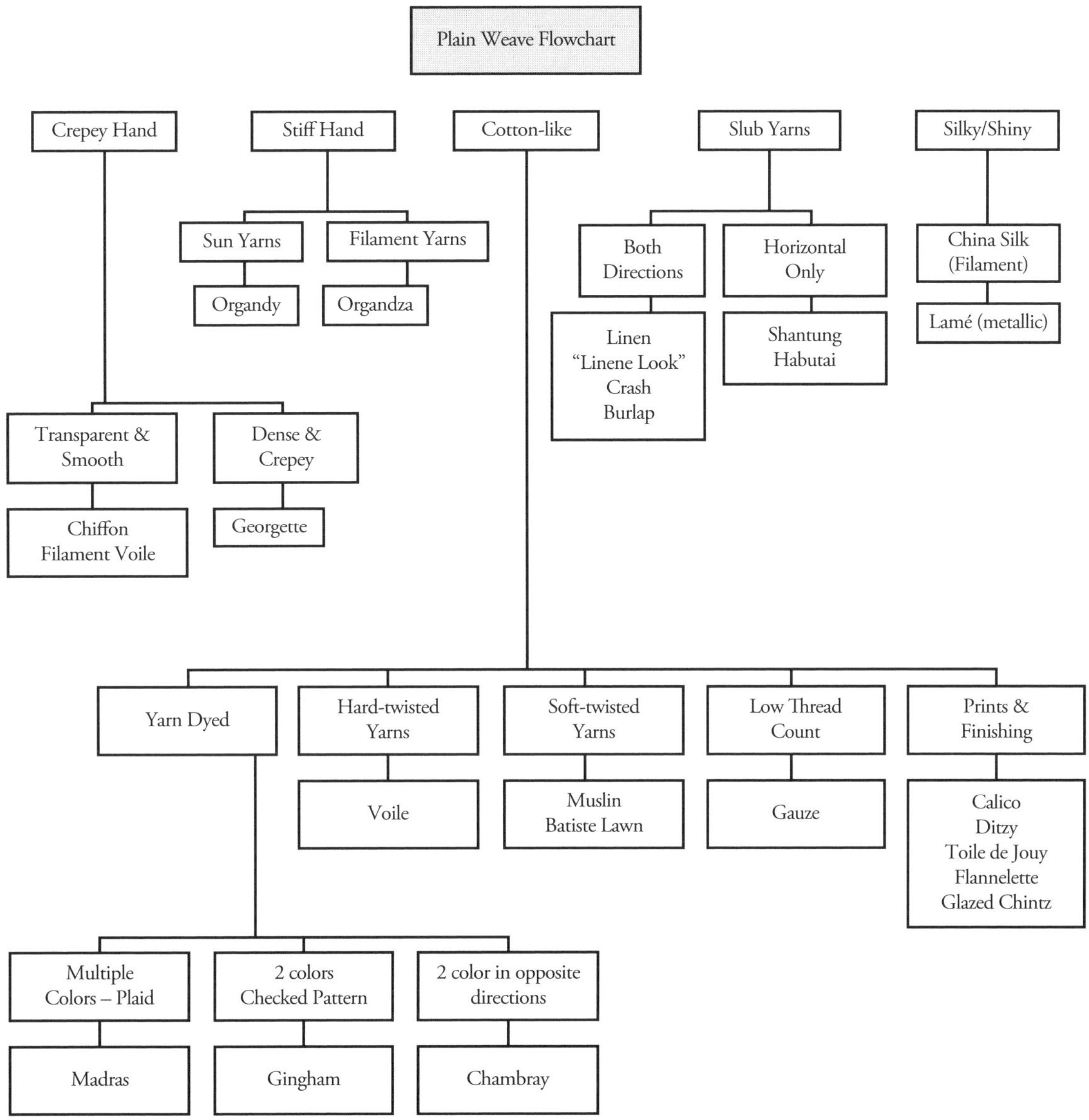

Figure 6.3 Plain-weave flowchart

Interior Design Focus: Plain Weaves

Window Treatments

Most plain weaves lend themselves well to curtains and sheers, due to their light weight and transparency. Sheers are best constructed of chiffon, georgette, voile, filament voile, organdy, and organza. Curtains (as opposed to drapes) can be constructed of most other plain weaves, such as gingham, linen, gauze, or chambray.

Sheets/Bedding

A densely woven plain weave, often called percale, is the most common fabric choice for sheeting. Sheets are best made of cotton with a high thread count due to cotton's absorbency, wicking, soft hand, and thermal conducting properties. A high thread count (400 to 1,000) will improve resilience and dimensional stability.

Today, sheets are often made of microfibers due to their luxurious, soft hand. Microfiber is often made of polyester, which is more durable with improved resilience and dimensional stability, but it is not absorbent. They are also thermal retainers, which is usually not desired in a sheet. Microfibers can offer very high thread counts, however. Check the labels; sheets have become more complicated.

"Cotton rich" is a term found in sheets, which means that it is mostly cotton, but is blended with polyester. Although this may reduce absorbency, it will also increase resilience and dimensional stability. Sheets may also be made of bamboo rayon, which offers superior absorbency, wicking, a soft hand, and is naturally antibacterial—all very desirable for sheets.

Upholstery

If the plain weave is heavy enough (above 6 or 7 ounces per square yard), it can be adapted to upholstery, due to the superior abrasion resistance found in plain weaves.

Activity 6.1 Fabric Evaluation by Weight

Name: ____ Date: ______________

We have discussed many different criteria that are used in fabric evaluation. One of the most important distinguishing characteristics of a fabric is weight. Fabric is often described as a "12-ounce velveteen" or "7-ounce jersey."

Objective: To evaluate and determine the weight of fabrics.

Procedure

1. Locate and prepare test specimens.
   a. Find two very different weights of upholstery fabrics: the lightest and heaviest that you can find (minimum two ounce difference).
   b. Find two very different weights of lightweight drapery sheers or lace (minimum two ounce difference).
   c. Your samples should not include seams, selvages, pinked edges, or hems.
   d. Cut one sample of each fabric to precisely 5" × 4". Inaccurate cutting yields inaccurate results.
2. Weigh test specimens
   a. Using a gram scale, tare the balance (clear the balance of the scale to zero).
   b. Weigh each sample and record the weight in grams. Make sure that the entire swatch is on the scale. (You may have to fold the fabric so that it fits on the scale.)
3. Compute ounces per square yard. __________ grams × 45.72 ÷ 20
   a. Multiply the results (grams) by 45.72 to convert the number to ounces.
   b. Because the swatch is 5 × 4 inches, divide the results by 20 to convert the measurement to a square yard (length by width).
   c. Express the result as ounces per square yard (oz./sq. yd.). Do not round; include two decimal points.
   d. Record your results in the worksheet below.
4. Mount the samples.
   a. Label the samples.
   b. Mount them on the following page with the warp in the vertical direction.

Worksheet for Activity 6.1 Fabric Evaluation by Weight

	Weight	Multiplied by 45.72	Divided by 20 = Ounces per square yard
Lightweight Upholstery	______ grams		__________ oz./sq. yd.
Heavyweight Upholstery	______ grams		__________ oz./sq. yd.
Lightweight Sheers	______ grams		__________ oz./sq. yd.
Heavier Curtain or Drape	______ grams		__________ oz./sq. yd.

Name: ______________________________ Date: ______________

Fabric Weight: ______________ ounces per square yard.

Fabric Weight: ______________ ounces per square yard.

Name: ______________________________ Date: ______________

Fabric Weight: ______________ ounces per square yard.

Fabric Weight: ______________ ounces per square yard.

CHAPTER SEVEN

Plain-weave Variations: Basket and Rib Weaves

Reference Swatches

46. Monk's Cloth | 47. Canvas | 48. Iridescent Taffeta
49. Crepe de Chine | 50. Broadcloth | 51. Repp | 52. Bengaline
53. Ottoman | 54. Pincord | 55. Dimity

Two fabric structures are closely related to plain weaves: basket weaves and rib weaves. Like plain weaves, basket and rib weaves have a ratio of 1:1 in weave structure (over one and under one). However, there are variations that cause some of these balanced structures to ultimately have an unbalanced thread count. Although these fabric types are often referred to as plain weaves, they fall into a different category because, for various reasons, they perform differently than their plain-weave counterparts.

Basket Weaves

Like plain weaves, basket-weave construction is perfectly balanced, but two or more yarns act together as one to create the structure. Basket weaves are often represented by multiples in the yarn, such as 2:2, 3:3, 4:4, and so on. See Figure 7.1 for an example of a 2:2 basket weave.

Because of this multiple-yarn configuration, basket-weave fabrics have fewer interlacings. This means, if the count is low, the fabric is less strong and more decorative. In addition, basket weaves are often less expensive to produce. In terms of performance properties, these weaves offer improved resilience and elasticity but often at the expense of strength and abrasion resistance. They are also more absorbent than plain weaves and therefore have reduced dimensional stability.

Due to the reduced interlacings, more yarns may be packed in tighter to make fabrics such as canvas, which is a very sturdy and strong basket weave. See Swatch 47.

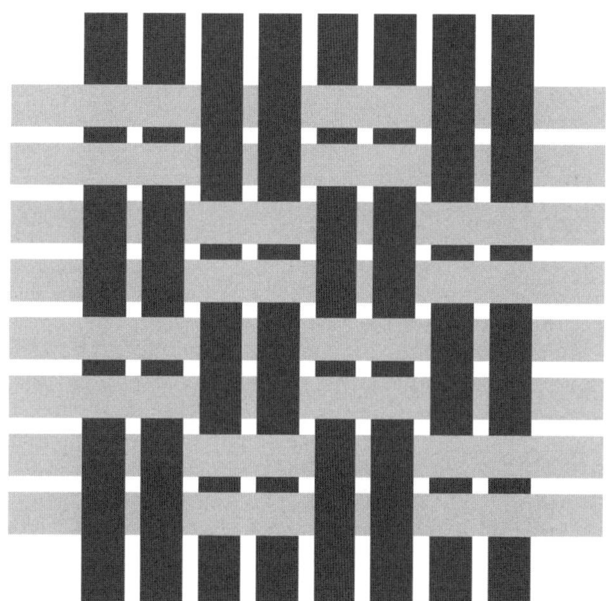

Figure 7.1 2:2 Basket Weave

A half-basket is made of two small threads to one larger thread, as seen in Swatch 124.

[Reference Swatches 46–47]

Rib Weaves

Rib weaves can be created in a couple of different ways. What is consistent is the lack of balance in the fabric: a pronounced rib is visible on the surface. Although this structure is still made in an over-one, under-one construction, there are either twice as many yarns in the warp direction as in the weft direction, or the weft yarns are thicker than the warp yarns. In either case, the rib is usually quite evident. The warp often completely covers the weft because the warp outnumbers the weft two to one. This results in a crosswise, or horizontal, rib.

Rib weaves produce a tightly woven fabric that offers strength, durability, and substantial body. On the other hand, the resulting fabric often does not drape well, and abrasion resistance can be a problem. Rib weaves are a good choice when structure and shape are the focus. Although most ribs are crosswise, lengthwise is also possible. A lengthwise rib generally improves drape, and a crosswise rib improves body.

[Reference Swatches 49–55]

Figure 7.2 may be used as a quick reference for identifying basket weaves and rib weaves, and Table 7.1 below compares the two weaves in terms of properties.

Table 7.1 Performance Expectations of Basket and Rib Weaves

Property	Basket Weaves	Rib Weaves
Durability	Low strength Better elasticity Poor abrasion resistance Poor dimensional stability	Better strength Poor elasticity Poor abrasion resistance
Comfort	More hydrophilic	Less hydrophilic than basket weaves
Care	Greater resilience	Less resilient than basket weaves
Appearance	More flexible	Better/more body Poor drape Better shape retention

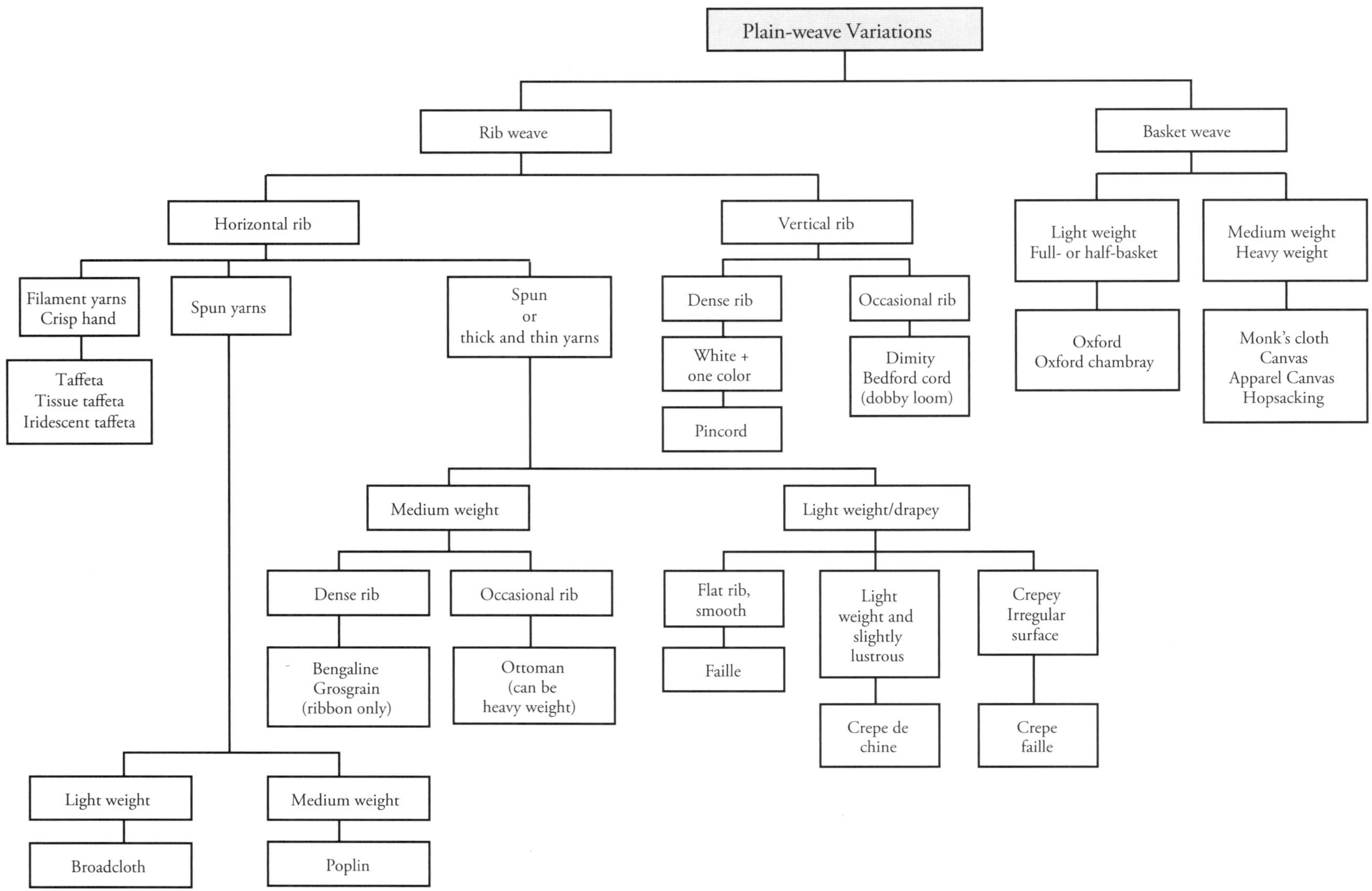

Figure 7.2 Fabric identification flowchart: Plain-weave variations

Activity 7.1 Woven Fabric Analysis

Name: ____ Date: ____________

Objective: To identify the components of a woven fabric and justify its use or application.

Materials required: Woven swatch provided by your instructor.

Procedure

1. Mount the swatch provided by your instructor, face up, with the warp placed vertically, attached only at the top.
2. Using your pick glass, knowledge from the burn test in Activity 4.1, and weight computations in Activity 6.1, analyze the swatch.
3. Record the following information:
   a. Fiber content: ____

      (e.g., 100 percent cotton)

      Yarn construction: ____

      (Spun or filament? Single or plied? If plied,how many?)

      Fabric name: ____

      (e.g., chambray, damask, tapestry)

      Fabric count: ____

      (e.g., 88x76)

      Weight: ____

      (expressed in oz. per sq. yd.)

CHAPTER EIGHT

Twill Weaves

Reference Swatches

56. Denim | 57. Drill | 58. Ticking
59. Herringbone | 60. Houndstooth | 61. Glen Plaid
62. Twill Plaid

The second of the three basic weaves is the twill weave. Twills are easily distinguished by the diagonal patterning. A twill's diagonals are called wales, which are actually raised ridges created by floats. A float is a skipped interlacing. If a yarn goes over two or more yarns, it is called a float. Twills are constructed with a minimum 2:1 ratio, (which means the weft floats over two warp yarns before going under one). Other common twills include 3:1 and 2:2. The basic pattern of interlacing (2:1, etc.) shifts to the right or left by one thread in each successive row, forming the diagonals (see Figure 8.1).

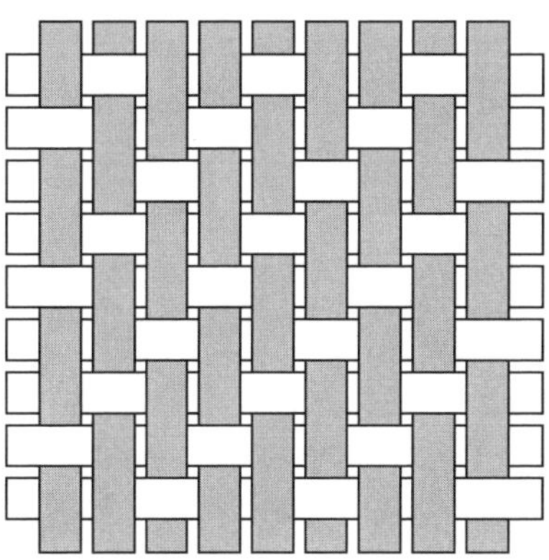

Figure 8.1 Twill weave

In general, twill weaves are denser, with a higher and more unbalanced thread count than plain weaves. The floats in a twill weave cause the yarns to interlace less often than plain weaves (which have a 1:1 ratio). For this reason, there are more yarns in the fabric, resulting in a fabric that is more compact and is generally stronger overall. Usually a heavier-weight fabric, twills are more durable and wear well, which is why twills are considered the workhorses of the three basic weaves. Twills are most often the best fabric choice where strength and durability is a consideration for high traffic areas in upholstery. Take this opportunity to look closely at your own denim jeans and the diagonal patterning.

A number of criteria can be measured in a twill weave, one of the most important being the direction of the diagonal. Twills are said to be right-handed (diagonal going up to the right) or left-handed (diagonal going up to the left) (see Figure 8.2). Traditionally, twills made of silk or wool were right-handed. Today, specific twill fabrics are likely to be right-handed, but that is no longer an absolute. Cotton fabrics were left-handed and are often still left-handed.

The degree of the angle of the diagonal is also important to consider. Most twills have a 45-degree angle, but there are also steep twills (63 degrees) and lazy or reclining twills (27 degrees) (see Figure 8.3). Generally, the steeper the twill, the higher the count and the more durable the fabric.

Figure 8.2 (a) Left-handed and (b) right-handed twill patterns

Twill Categories

Twills can be even sided (front and back are the same) or uneven (front and back are quite different).

Uneven Twills: 2:1 (1:2) or 3:1 (1:3)

Uneven twills have a minimum ratio of 1:2 or 2:1. This means the structure is woven in an over one and under two pattern. Because the ratios are unbalanced, the front and back of these fabrics are *always* different. The front (technical face) has the most pronounced wales and is more attractive and durable. Denim is an example of this type of twill; this fabric is easily recognizable because the front and back are remarkably different.

This category includes 3:1 or 1:3 as well. A 3:1 ratio is more different, front to back, than a 2:1 due to the longer floats. Notice the larger disparity between the black and white spaces (Figure 8.4).

[Reference Swatches 56–58]

Even-sided Twills

Even-sided twills are perfectly balanced, with a two-up, two-down structure. This makes it difficult to tell the difference between the technical face and back of these fabrics: they appear identical. Notice the even spacing between the black and white spaces in the figure (Figure 8.5). The patterning still moves to the right or left by one thread per row to create the diagonal appearance. Also note that if an even-sided twill is right-handed on the front of the fabric, it will be left-handed on the back.

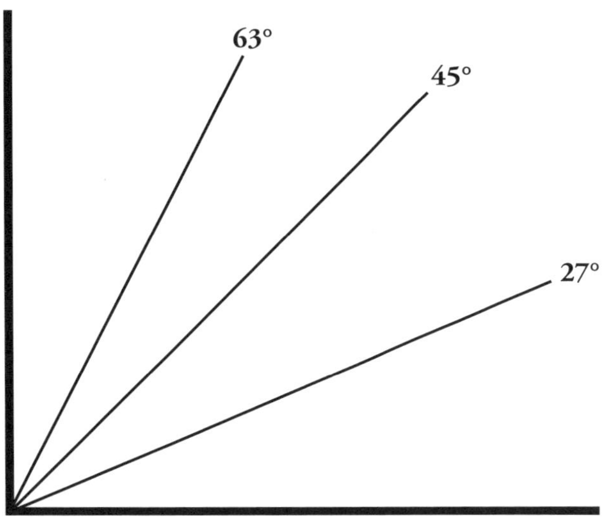

Figure 8.3 Twill degrees of diagonal

3:1 Twill pattern

Figure 8.4 Uneven twills: 3:1 left-handed twill

Figure 8.5 2:2 even-sided twill pattern

There is generally more warp than weft in twill fabrics, making it easy to identify which is which by pulling yarns out of each side of the swatch. Once you have pulled enough yarns to create a fringe in both warp and weft directions, check to see which direction has more yarns. The side with the most yarns is generally the warp.

Performance Expectations of Twill Weaves

Floats are longer in a twill (than in a plain weave), and this impacts the fabric in the following ways:

Durability

- Strongest overall weave
- Average abrasion resistance (reduced by float length)
- Improved elasticity
- Improved flexibility
- Higher, unbalanced count

Comfort

- Improved air and water resistance due to floats

Care

- Improved resilience

Appearance

- Hides soil
- Improved drape
- Improved luster

Refer to Figure 8.6, Fabric identification flowchart, as a quick reference for basic twill structures.

[Reference Swatches 59–62]

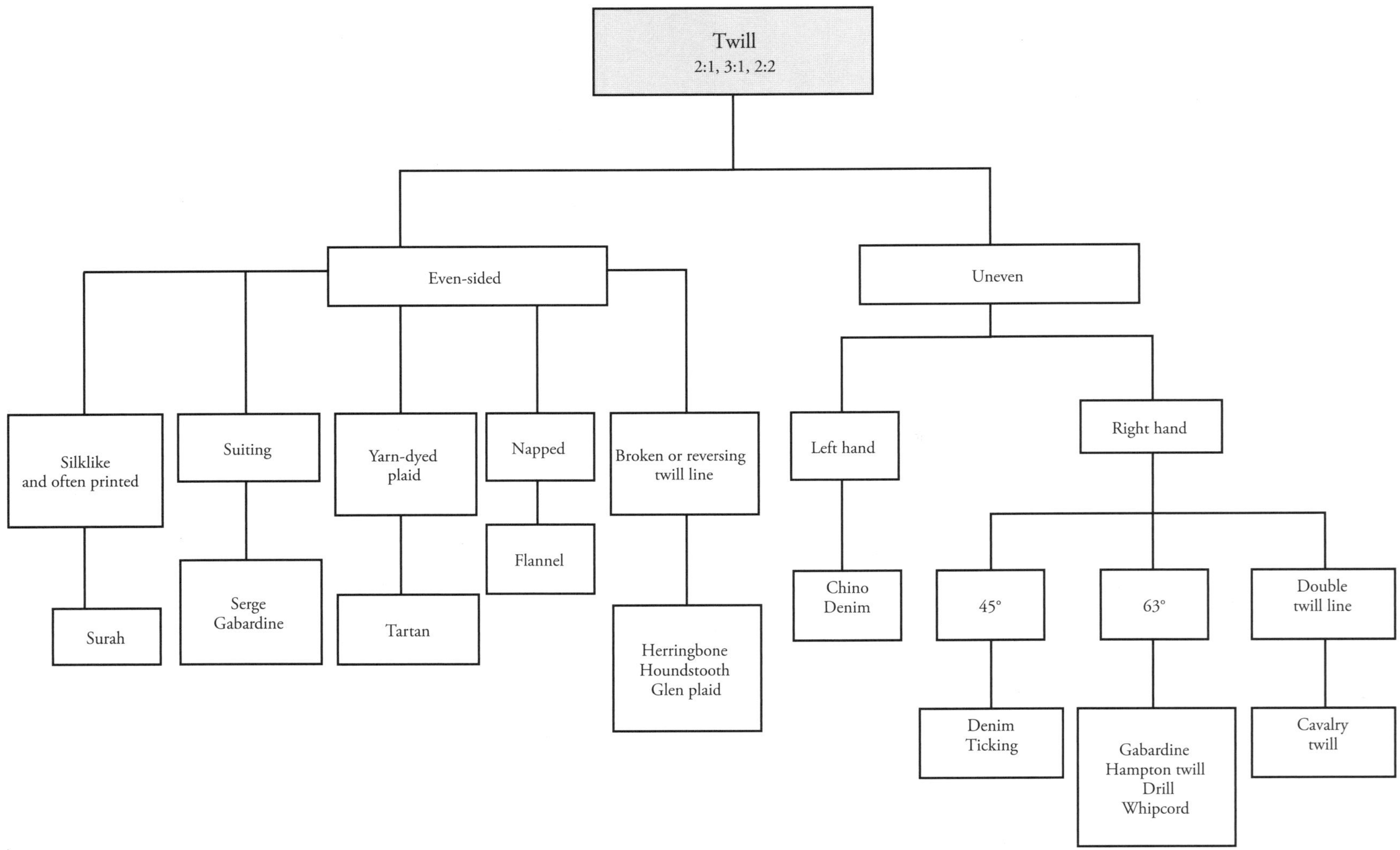

Figure 8.6 Fabric identification flowchart: Twill

Activity 8.1 Twill Application

Name: ______________________________ Date: ______________

Search online for a silhouette of a sofa or upholstered piece of furniture that might expect high traffic. Using the swatch book as reference only, find an example of a twill fabric that would be an appropriate choice to use as the fabric for this piece. Determine the facts of this fabric using skills already acquired. Include the sketch and swatch below, along with all pertinent information about the selection and placement. Acquire enough fabric to properly test it.

Sketch: Swatch:

Fiber content: ______________________________

Burn to identify

Yarn construction: ______________________________

Spun or filament

Fabric structure: ______________________________

This should be a twill weave.

Fabric name: ______________________________

Exactly which twill is this? Use your swatch book to verify.

Count: ______________________________

Example 68 × 42.

Weight: ______________________________

Show the math.

Name: ________________________________ Date: ______________

Finally, in a brief sentence, justify your selection of fiber and fabric for the chosen silhouette.

Fiber ______________________ is a good choice because ______________________________

____

____

Fabric ______________________ is a good choice because ______________________________

____

____

CHAPTER NINE

Satin Weaves

Reference Swatches

63. Satin | 64. Crushed Satin | 65. Antique Satin
66. Flannel-back Satin | 67. Sateen

Satin is the third and last of the three basic weaves. One of the main characteristics of this weave is the length of the floats: satin has longer floats than twill. Satin ratios begin at 4:1 (or 1:4) and can range as high as 11:1. This means that the fabric is not balanced; it is either warp- or weft-faced (depending on the direction of the floats). Because the ratio is so unbalanced, the thread count is likewise unbalanced and usually quite high. This creates a distinctly different face and back. *All satins have a front and back that are remarkably different from each other.*

Traditionally, satins are warp-faced (floats are in the vertical direction), which produces maximum luster. Sateens have typically been weft-faced (floats in the horizontal direction). The industry is changing rapidly, and these rules do not always apply; there are many weft-faced satins and warp-faced sateens.

For this reason, the definitions of both have been simplified. What seems to be always true is that sateens are made with spun yarns, and satins are made with filament yarns, often with little or no twist. In terms of performance, look to the satins for smoothness, luster, and drape. Abrasion resistance is compromised, and seam slippage (a shifting of slippery yarns that causes the fabric to open near a seam) can be a problem with these weaves.

Satin fabrics unravel and snag easily, but because of their high and unbalanced count, they exhibit surprisingly good strength. They can therefore be a good choice for upholstery due to weight and density. Satins and sateens can also be an excellent choice for luxurious, high quality sheets. In both cases, a high thread count is paramount to maximum performance.

Although satins generally have a higher count, the relatively lower counts in some satin fabrics are detrimental to overall expected performance, causing the yarns to shift and the fabric to be even less durable. A satin with a low count that does not hold its shape is known as a sleazy fabric. See Figure 9.1.

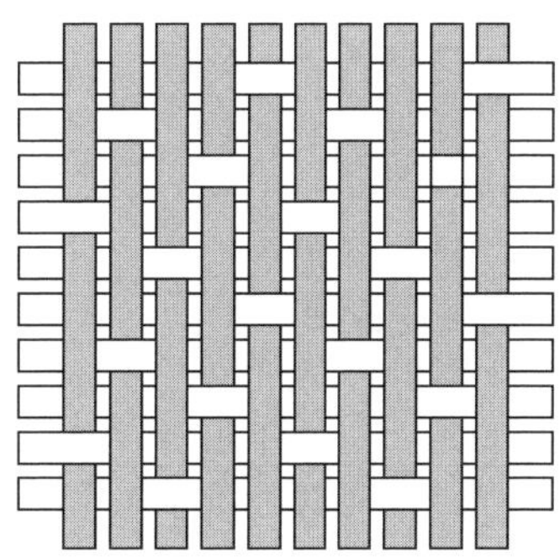

Figure 9.1 Satin weave

Performance Expectations of Satin Weaves

The following list of characteristics applies to satin weaves:

Durability

- Strength
- Reduced abrasion resistance
- Excellent flexibility
- Easily snags
- Unravels the most

Comfort

- Good wind resistance
- Good thermal retention

Care

- Good resilience

Appearance

- Great luster
- Smooth, silky hand
- Wonderful drape
- Seam slippage can be a problem

Refer to Figure 9.2 as a quick reference for identifying satin weaves.

At this time, it will be helpful to stop to review the three basic weaves. Figure 9.3 presents the weaves on a continuum, running from the fabric with the shortest floats, plain weave, to twills, and then finally to satins. Notice that as floats get longer, aesthetic properties (resiliency, elasticity, luster, and drape) improve, but durability (strength, abrasion resistance, dimensional stability, and raveling) generally decreases. Conversely, as floats get shorter, aesthetic properties generally suffer, but durability improves. Twills sit in between plain weaves and satin weaves in terms of float length. The shorter floats give the fabric just the right balance of durability properties, causing them to have the best overall strength. When you factor in their excellent soil resistance, twills are perfectly appropriate for daily use in applications where longevity is a consideration. Table 9.1 offers a comparison of the three basic weaves in terms of properties.

Study Guide

Table 9.2 offers a quick reference guide to assist you in identifying fabrics. You will find Table 9.2, Quick Reference Guide for Three Basic Weaves, preceding the swatch boards. Please remove the table and place within Chapter 9 for easy referencing.

[Reference Swatches 63–67]

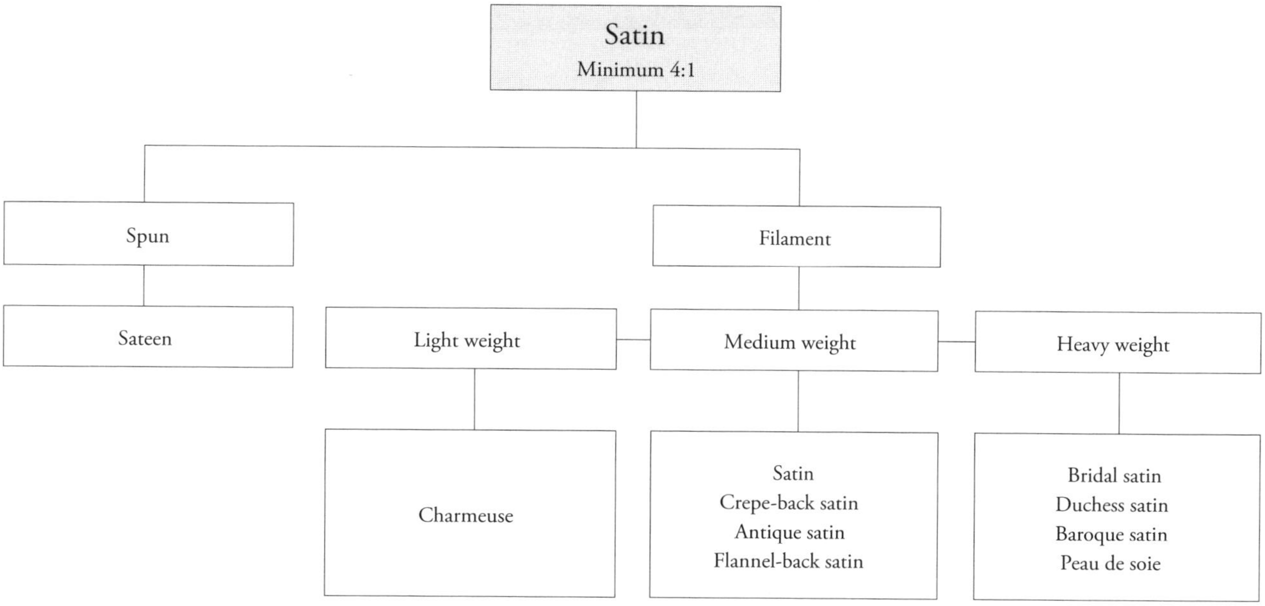

Figure 9.2 Fabric identification flowchart: Satin

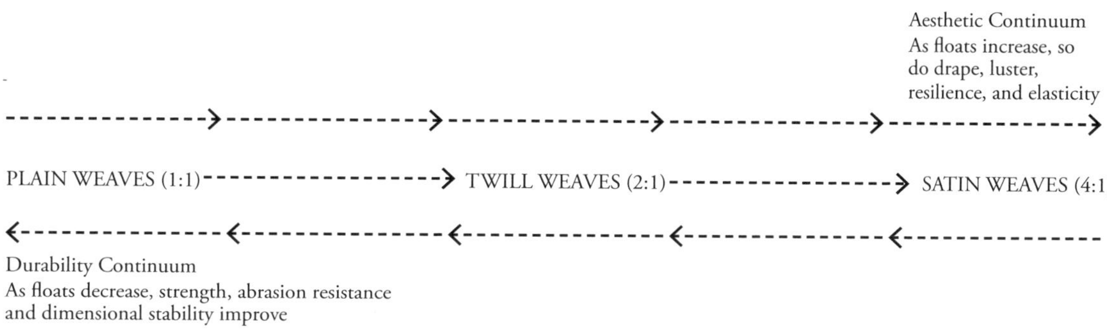

Figure 9.3 Aesthetic and durability continuum for plain, twill, and satin weaves

Table 9.1 Comparison of Basic Weaves

	Plain Weaves	Twill Weaves	Satin Weaves
Structure ratio	1:1	2:1, 3:1, 2:2	4:1–11:1
Front/back contrast	Front and back are always the same. (Front and back may initially appear different due to finishes or prints.)	2:1 and 3:1: Front and back are different; 2:2: Front and back are the same.	Front and back are always different.
Abrasion resistance	Excellent abrasion resistance	Moderate abrasion resistance, some fibers may get shiny with excessive wear	Poor abrasion resistance
Strength	High tensile strength, low tearing strength	Strongest overall weave, tearing, and tensile strength	High tearing strength, low tensile strength
Thread count	Lowest, most balanced thread count with most interlacings and no floats	Higher, less balanced thread count with reduced interlacings and more floats	Highest, most unbalanced thread count; fewest interlacings and longest floats
Snagging/raveling	Not prone to raveling or snagging	Very little raveling and snagging	Snagging and raveling can be a problem
Soiling	Easily soiled, easily cleaned; soil is easy to see	Soil resistant, harder to clean, soil hides better	Difficult to soil, but easy to see soiling because of luster
Thermal properties	Most breathable, especially basket weaves	Better thermal retention	Offers wind and water resistance
Weight	Lightest weight, can be sheer	Can range from curtain to upholstery	Usually upholstery weights, but can be used for windows
Resiliency	Poor resilience	Better resilience	Excellent resilience
Elasticity	Poor elasticity	Improved elasticity	Excellent elasticity
Drape	Poor drape, good body	Good flexibility and drape	Excellent drape
Luster	No luster	Some luster with surah, but most have little luster	Most lustrous
Cost	Cheapest fabric to make	Increased expense	Most expensive to make

Activity 9.1 Weave Comparison Graph

Name: ______________________________ Date: ____________

Part a

Use the graphs below and on the following page to chart out (1) a plain-weave structure, (2) a 2:1 twill-weave structure, and (3) a 4:1 satin-weave structure.

Part b

Collect two different magazine tear-sheets with two different images (and color combinations). One tear-sheet will represent warp; the other, weft. Slice the pages into ¼-inch to ½-inch strips and weave them into the structures you charted in Part a. Note how your eye mixes the color and pattern into a woven structure. Finish the edges neatly.

Plain Weave

Activity 9.1 Weave Comparison Graph *(continued)*

Name: ______________________________ Date: ____________

Twill Weave

Satin Weave

CHAPTER TEN

Complex Weaves

Reference Swatches

68. Seersucker | 69. Dobby—Filament | 70. Dobby—Spun
71. Bird's-eye Piqué | 72. Waffle Cloth | 73. Goose-eye Twill
74. Satin Stripe | 75. Crepe/Momie Weave | 76. Extra-yarn Weave
77. Clip Spot | 78. Dotted Swiss | 79. Tapestry | 80. Cotton Damask
81. Printed Damask | 82. Flamestitch | 83. Brocade
84. Brocatelle | 85. Liseré | 86. Double Weave
87. Pocket Cloth | 88. Cloqué | 89. Matelassé | 90. Marquisette

Complex weaves are the stars of the interior designer's toolkit. They are loaded with personality and can be quite seductive to the client. An entire room concept can be built around a single, well-placed brocade or other jacquard fabric.

Complex weave is a generic term used for fabrics that are made of a combination of the basic weaves discussed in the previous chapters. They can also be made of many colors of yarns in a single weave. They are, in fact, complicated. These weave structures are made on special looms or loom attachments, such as a dobby or a jacquard, that allow detailed manipulation of the design. Complex weaves are more expensive than the simpler fabrics referenced in the preceding chapters, but they are generally heavier, more durable, and wear better for increased longevity.

Complex weaves are not, in and of themselves, a specific weave type, like twill or satin. They are a combination of fabric structures and variations. The following pages and swatches illustrate some of the many fabric types in this category (Figure 10.1).

Slack-tension Weaves

Seersucker is made with an alternating pattern of tight and loose (slack) tension of the warp threads. This causes a vertical puckering in the fabric. Usually made of cotton, this fabric has a wonderful summer appearance that, due to its intentional puckering, does not require ironing. The fabric is often characterized by vertical stripes and can be used for light weight curtains, home décor accents, and bedding accessories.

[Reference Swatch 68]

Dobby Weaves

Dobby is the name of a loom type or attachment that weaves fabric with small, geometric designs, such as diamonds. This is a simple structural design, made with two different weaves that create a foreground and background in the design. The scale of the design is relatively small and simple. Dobby designs include piqués, waffles, momies, crepes, honeycombs, and twill variations.

[Reference Swatches 69–75]

An Overview of Crepes

Crepe is a general term for a group of fabrics that have a crepey or pebbly surface. Crepe does not relate to a single fabric structure. It is made using one of several methods of construction. The overview that follows reveals some of the differences in construction. There are many other crepe type fabrics besides the ones presented in this text. As you will see, structure is not a constant in this category.

Crepe Fabrics

- Crepe de chine (Swatch 49): Rib weave made with alternating S & Z–twisted yarns in the weft direction
- Crepe-back satin: Satin weave made with a combination of untwisted yarns in the warp and over-twisted yarns in the weft
- Crepe or momie weave (Swatch 75): Made with a momie weave on a dobby loom

The constant in these fabrics is performance, hand, and appearance, not structure.

A crepe fabric offers the following:

- Excellent drape, resiliency, and elasticity
- Poor luster; a textured, or rough, hand; and poor strength

Crepes can be created in the following ways:

- Weave structure: As in a momie weave made on a dobby loom
- Yarn structure: Over-twisted, S & Z twist, or textured filament
- Finishes, such as embossing

Extra-yarn Weaves (aka Supplemental Warp or Weft)

An extra set of yarns (beyond the warp and weft) are used to create isolated structural motifs. When the extra yarns are not creating the design, they float along the back of the fabric until needed again. As these fabrics are subsequently nailed to a frame for upholstery, the floats are not a problem. If these floats are subsequently cut off, the fabric is called a clip spot (clipped spot, clip dot, dot spot).

[Reference Swatches 76–78]

Jacquard Weaves

Jacquard weaves are the most complex structures of all, often incorporating detailed, pictorial imagery. These fabrics are generally the heaviest weight, quite suitable for upholstery. A woven fabric is a grid, with two sets of yarns at right angles to one another. There is nothing curvilinear in a woven structure. Yet jacquard fabrics exhibit complex imagery, using many rounded edges. A Jacquard loom makes this possible, using so many tiny threads and individual manipulations to realize the design. Think of a computer monitor, which has the same grid-like limitation. With enough pixels, however, one begins to see round structures. Jacquard technology works in much the same way: If there are enough threads, smooth, fluid curves can be achieved.

Jacquard is the name of a loom, not a fabric, although fabrics made on a Jacquard loom are often simply called jacquards. Since our purpose is to give each fabric in this text its own unique identity, we must name each fabric according to structure and nuance.

The Jacquard loom was invented in 1801 by Joseph-Marie Jacquard in Lyon, France, to replace incredibly arduous weaving techniques that often resulted in ten-plus years of toil to create a single fabric. The first Jacquard looms operated on a punch-card system, much like a player piano. These looms adeptly replaced the Chinese draw-loom (and the draw-boy, who lifted the threads, like a puppet master) for making complex structures. Today Jacquard looms have been computerized. Indeed, anything that you can draw or scan into a computer can be woven on a Jacquard loom. These electronic jacquards have become today's standard, working at very high speeds.

Characteristics of Jacquard Fabrics

- Each thread is individually controlled without harnesses.
- Multiple weaves in the same fabric are possible.
- One weave with many colors is possible.
- Any image can be achieved.

Jacquard technology often produces larger, heavier-figured designs. The fabrics in Swatches 79–85 were named for fabric structures that predate the Jacquard loom. Fabrics are distinguished from one another according to the processes that are used to create them. Use Figure 10.1 as a handy reference to help you identify complex weaves. See the individual swatch boards to determine the differences between damasks, brocades, and tapestries.

[Reference Swatches 79–85]

Double Weaves

Double weaves are essentially two fabrics that are woven together. In this way, the front and back of the fabric need not have a relationship to each other; they can be remarkably different. Double weaves can also be the same on the front and back, as in a double-sided satin. Sometimes a double weave can be mistaken for a damask, as the two sides of the fabric may reverse. Check the edges to determine the number of sets of yarns. Often heavyweight, they can have as many as five sets of yarns (two warps, two wefts, and a fifth yarn to weave the fabrics together). This makes the fabric more durable and suitable for upholstery. Matelassé is a double cloth with a three-dimensional surface that is prevalent in the interior market for upholstery.

[Reference Swatches 86–88]

Leno Weaves

A leno weave is often used for casement, lacy, and sheer structures. Two warp yarns are twisted around the weft yarns in a figure-8 configuration. This provides an open and transparent weave structure that is stable and strong, such as the swatch provided here, marquisette.

[Reference Swatches 89]

Interior Design Focus: Complex Weaves

Window Treatments

Complex weaves are often used for heavier drapery. Fabrics such as damask, liseré, and brocatelle are often used to accessorize a room in this way. Lighter fabrics, such as dotted swiss, lend themselves well to lighter weight sheers.

Sheets/Bedding

Damask has long been a staple of elegant cotton sheeting. Brocades and matelassés are often more appropriate for the top of the bed and home décor accessories.

Upholstery

Upholstered fabric is often made of complex weaves, due to its heavier weight, appearance, and patterning. Jacquards offer superior abrasion resistance compared to a printed chintz, for example.

Refer to Figure 10.1 as a quick reference for identification.

Study Guide

Refer to Table 16.1 for a list of fabrics organized by final applications. This will assist you in making appropriate fabric choices for end use.

Table 10.1 offers a quick reference guide to assist in identifying complex and pile weaves. You will find Table 10.1, Quick Reference Guide for Complex and Pile Weaves, preceding the swatch boards. Please remove the table and place it within Chapter 10 for easy reference.

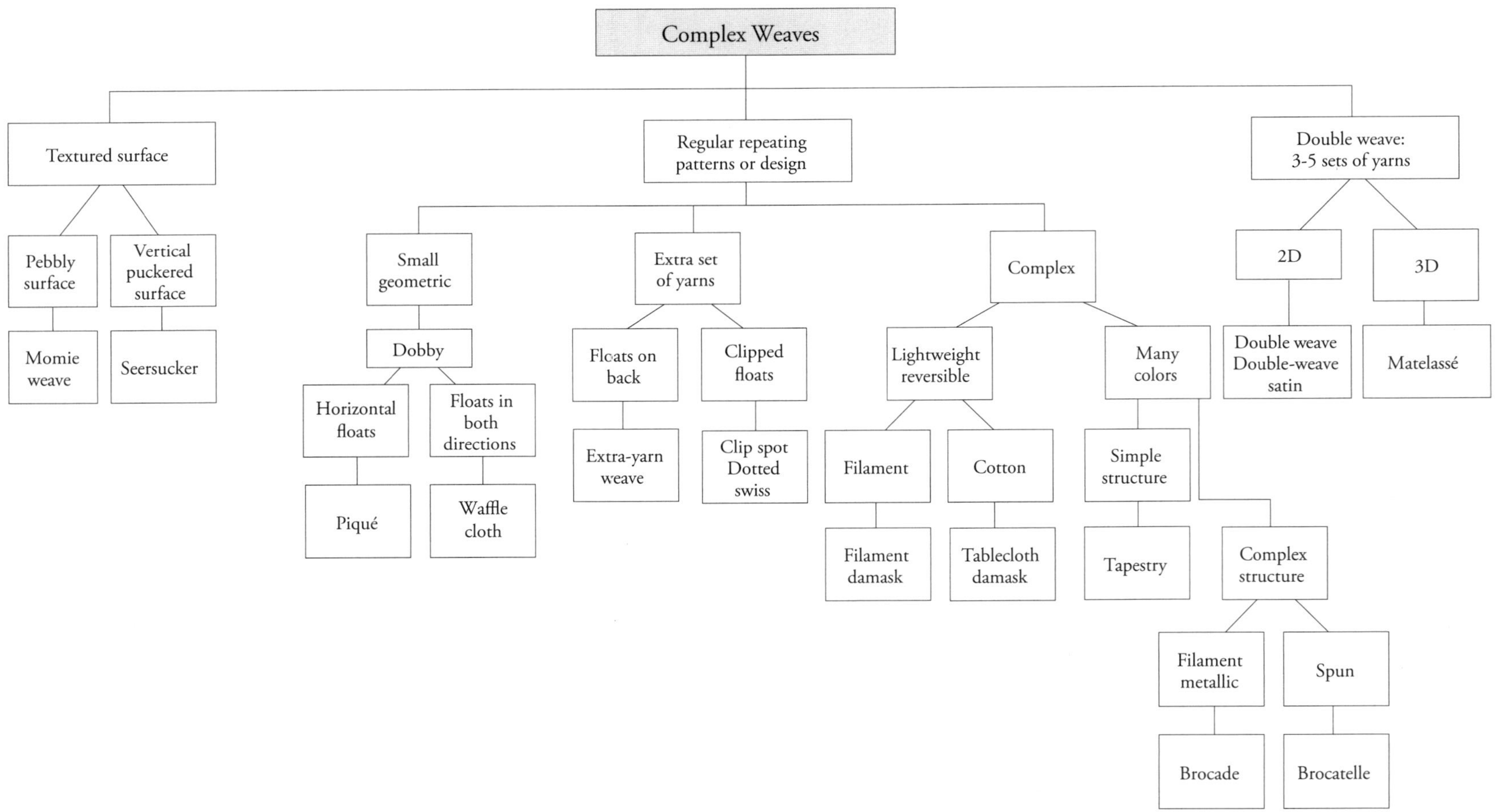

Figure 10.1 Fabric identification flowchart—Complex weaves

Activity 10.1 Research Project: Storybook

Name: ______________________________ Date: __________

Objective: To understand the correlation of fabric selection to application.

Procedure

1. Find two images: one of an upholstered piece of furniture and one of a sheer window treatment. These should both coordinate into a single room.
2. Find and identify two different fabric alternatives for each piece listed above. Consider fiber and fabric construction and relevant properties. Each fabric should be different in structure, not just color. Use Table 16.1 to assist you in fabric selection.
3. Identify each swatch by fiber content, yarn construction, fabric name, and weight.
4. Choose the fabric that is the best choice for your design. Mark it as your recommendation.
5. For the swatch chosen in Step 4, write three performance properties that justify why this fiber/fabric combination is the best choice for the product. (Example: Hydrophobic. Polyester is hydrophobic, which means it is not absorbent. This is an excellent choice for swimwear as it will dry quickly.)

Professionalism

- Present the project professionally in a flat folder.
- Professionally cut the fabric swatches to 3 × 4 inches.
- Mount swatches with the technical face up and the warp direction vertically oriented. Using double-stick tape, attach only the top edge of the swatch.
- Use typed labels to identify fiber content, fabric name, yarn construction, and weight.
- The layout of the project must be neat and professional.

Notes

- Avoid negative or neutral properties (e.g., poor strength, average drapability).
- Properties should apply to the fabric as well as the fiber.
- Use formal names of properties, followed by a definition in your own words, and explain how this property improves or enhances the garment.
- Use the swatch book as a reference, not a source. Do not use swatches from the swatch book.
- Use only traditional fabrics (with a name) that can be researched and identified.
- Each swatch must have a different woven fabric name. There should be no duplicates.
- Select fabrics that are 100 percent of a fiber. Avoid blends.
- Do not use denim, leather, vinyl, plastic, fur, or lace.

CHAPTER ELEVEN

Pile Weaves

Reference Swatches

91. Terry Cloth | 92. Widewale Corduroy | 93. Velveteen
94. Velvet | 95. Panné Velvet | 96. Voided Velvet
97. Grospoint

The English word pile is from the Latin word meaning hair, and the fabrics in this category are all hairy. Even more important, these fabrics are three-dimensional. Pile fabrics can be used for everything from bedding and towels to carpeting and stuffed animals. These fabrics can be woven, knitted, or tufted. The focus of this chapter will be woven pile fabrics.

Construction of Pile Weaves

In the construction of a pile weave, there are three sets of yarns: a vertical set (warp), horizontal threads (weft), and a set of yarns making perpendicular loops on the surfaces of the other two, creating a three-dimensional structure. The warp and weft interlace and form a tightly woven ground weave. For all of the fabrics presented in this chapter (except terry cloth), the perpendicular loops are later sheared and brushed in the finishing process (see Figure 11.1).

Once sheared, the nap, or pile, of the fabric stands up, but it is often not perfectly perpendicular to the ground weave. The nap is often at a slight angle. When laying the fabric out, it is imperative that all the pieces be laid in the same direction; otherwise, the light will reflect the different directions of the pile, causing the pieces to look as though they do not match. For this reason, more fabric is required for construction with pile fabrics.

Although most pile structures have a plain-weave ground, some are constructed on a twill ground, which makes the fabric stronger and more durable. As always, the higher the count of the basic ground weave, the more durable the fabric will be. A higher-count pile weave will also retain its shape and resist crushing far better than a lower-count alternative.

Consider carpeting, which is a pile structure that is large and three-dimensional. If the count of the carpeting is low, your footprints might be evident as you

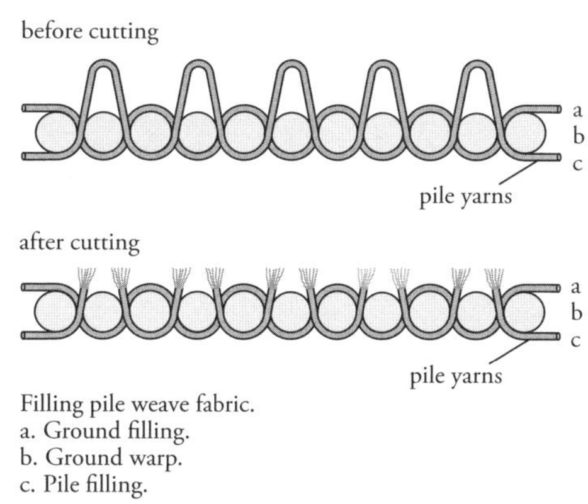

Figure 11.1 Pile weave before and after cutting

walk across the carpet. With a higher-count carpet, footprints are not left behind as readily, since the pile is dense enough to avoid laying down. Carpets are covered in more detail in Chapter 14. Now consider a pile weave laid on a sofa. More texture will be created with a lower count. A higher-count fabric will have better compressional resilience (will recover better from bending or crushing).

Generally, pile weaves are denser, heavier, less pliable, and less drapable than any of the two-dimensional fabrics, but this is not always the case. If the ground weave is a chiffon, for example, the resulting fabric will be an exceedingly lovely and drapable velvet called chiffon velvet. This could be an excellent choice for a window treatment. The properties that follow are generalizations that can, once again, be mitigated by myriad factors.

Performance Expectations of Pile Weaves

Durability

- Good strength
- Moderate abrasion resistance

Comfort

- Thermal Retention
- Absorbency: Hydrophilic

Care

- Good resilience

Appearance

- Poor drape
- Fuzzy hand
- Can have a high luster

[Reference Swatches 91–97]

Study Guide

Table 10.1 offers a quick reference guide to assist in identifying pile and complex weaves. You will find Table 10.1, Quick Reference Guide for Complex and Pile Weaves, preceding the swatch boards. Please remove the table and place within Chapter 10 or 11 for easy reference.

Activity 11.1 Fabric Structures

Name: ______________________ Date: __________

Attach a 3 × 4-inch swatch of each weave construction listed below, technical face up, with the warp in the vertical direction. Do not use swatch book swatches. Do not share swatches with another student.

Plain Weave—Name: ______________	Twill Weave—Name: ______________
Satin Weave—Name: ______________	Pile Weave—Name: ______________
Dobby or Extra-yarn Weave—Name: ______________	Jacquard Weave—Name: ______________

CHAPTER TWELVE

Weft Knits

Reference Swatches

98. Jersey | 99. Double-knit Mattress Pad | 100. Sliver Knit
101. Fleece | 102. Bonded & Embossed Pile Knit

Knitting is the second major method of constructing a fabric. Weaving was invented thousands of years before knitting; knits are relative newcomers. As such, knitting structures borrow terms and even structures from woven fabrics. Some knit fabrics even share the same or similar names as woven fabrics; it can be quite confusing.

As knits provide superior drape, elastic recovery, and resilience, they easily find a place in the home market. They offer a distinctive texture, appreciated in home décor. Knit fabrics can be found in bedding (afghans and other blankets), window treatments, accent pillows, throws, and small upholstered pieces.

Knits are made with a series of loops, imparting more elasticity than a woven fabric. In the past decade, the knit market has grown exponentially in the apparel industry, taking on a much larger share of the overall textile marketplace. In determining a fabric's structure, the main difference between woven and knit fabrics is that woven fabrics are made with two sets of interlacing yarns, whereas knits are constructed with one set of interlooping yarns. Together, Figure 12.1 and Table 12.1 provide a comparison of the two basic fabric constructions, including differences in performance expectations. For the purposes of like comparison, assume all other variables (fiber content,

Basic woven fabric

Basic knit fabric

Figure 12.1 Comparison of (a) woven and (b) knit fabrics

Table 12.1 Comparison of Weaves and Knits

When using this chart, consider two fabrics that are made of the same fiber, yarns, and the same basic weight.

Knit Fabrics	Woven Fabrics
Interloopings	Interlacings; yarns at right angles to each other
Visual: Looks like Vs	Visual: Looks like a grid
One set of yarn	Two sets of yarns
Difficult to unravel; unravels from top or bottom only	Easy to unravel from four sides
Better resilience, recovers from wrinkling	Better shape retention
More elasticity	Better dimensional stability
Thermal retainer	Better cover
More porous (air permeability), warmer in still air	Better insulation and heat retention in moving air
Uses more material	Lighter weight
Prone to snagging	Good abrasion resistance
Runs, ravels	Prone to seam slippage
Better drape and flexibility	More body
More hydrophilic	Less absorbent
Faster production technique	

weight, and yarn construction) are equal; analyze the differences in structure only.

Note in Figure 12.2 that the warp and weft directions of woven fabrics have become the wales and courses, respectively, of knit fabrics. As in woven fabrics, the measure of the wales and courses are often important determinants of the quality of the fabric.

Knit counts are different from woven counts. Woven fabrics have two sets of yarns, and each yarn is counted. Since an entire fabric can be knitted with a single yarn, counting yarns becomes irrelevant. In a knit fabric, the stitches (Vs) are counted across (for a count of wales) and down (for a count of courses).

A greater number of wales per inch will create a fabric that has more stability and less shrinkage in width. A greater number of courses will allow for less shrinkage in length. Overall, a higher stitch count will increase the elasticity (ability to recover from stretching) of a fabric.

Fabric Gauge

Although both wales and courses are evaluated by the discerning student, the industry often refers to knits by gauge. Gauge is the number of wales per inch, and a fabric is referenced as, for example, a 16-gauge jersey. Higher numbers mean that there are more stitches per inch, which means that the stitches are smaller and the fabric is finer. A bulky knit throw could be 8–10 gauge, while jersey sheets are easily 30–40 gauge.

Evaluating Stretch

When discussing knits in general, a discussion of stretch is essential. The quality of elasticity should be differentiated from basic stretch. Elasticity is the ability of a fabric to return to its original dimensions after stretching. Stretch, on the other hand, is the amount that a fiber or fabric will elongate. For interior applications only moderate and stable knits should be considered. Stable knits have little actual stretchiness; some have none at all, but will still recover nicely after being stretched or stressed, and could be used for lace window treatments and some upholstery. Jersey is a common knit with moderate stretch, which means it stretches and recovers up to 50 percent of its original dimensions. This fabric is used in sheets, blankets, and fleeces.

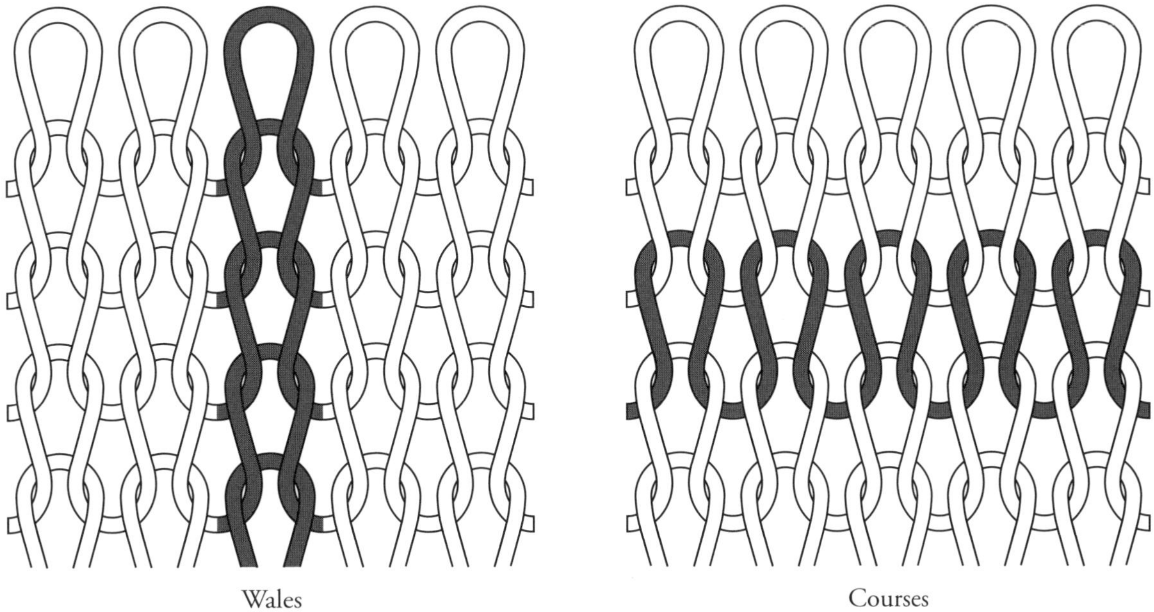

Figure 12.2 (a) Wales and (b) courses

Knit Quality Criteria

When purchasing knits, the following criteria should be evaluated:

- Size of fiber: Larger fibers pill less than smaller fibers. Larger fibers are shaggier and hairier than small fibers.
- Yarn construction: Spun or filament. Spun will pill but is soft and cozy; filament has more luster and drape.
- Amount of yarn twist: Higher twist means that the fabric will wear better, drape better, and pill less.
- Yarn type: A plied yarn is stronger than a single.
- Direction and degree of stretch: Consider the various amounts of stretch for different uses.
- Yarn count: A higher count means the garment will pill less, shrink less, and wear better.
- Gauge: This characteristic addresses the number of stitches per inch, which contributes directly to the shrinkage, drape, and resilience of the fabric. Even with jersey sheets, count matters.
- Finishing: Finishing processes can make a difference in terms of performance; the fabric should be properly steamed and pressed to avoid glazed or shiny areas.

The Four Basic Knit Stitches

Knit fabrics are made with four basic knit stitches. Recognizing them will help you to identify fabrics (Figures 12.3–12.5).

1. Knit stitches (Figure 12.3a) have wales on the front of the fabric and courses on the back. Wales are vertically oriented; courses are horizontal. This stitch is the foundation of jersey. Refer to Swatch 113, whose front has wales and back has courses.
2. Purl stitches (Figure 12.3b) are the opposite of knit stitches. They have courses on the front and wales on the back.
3. A miss stitch (Figure 12.4) is a missed stitch, also known as a float stitch. Miss stitches are often used to create color effects. They reduce fabric weight as well as the stretch and width of the fabric. One color of the yarn floats along the back while another color is being knitted on the face of the fabric.
4. A tuck stitch (Figure 12.5) is an elongated skipped stitch that creates a hole in the fabric. This stitch is used to create lacy, textural design effects. It also increases fabric weight, width, and thickness.

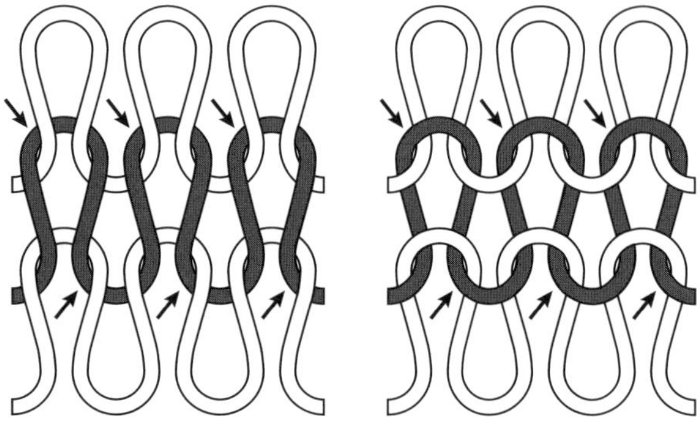

Figure 12.3 (a) Knit stitch and (b) purl stitch

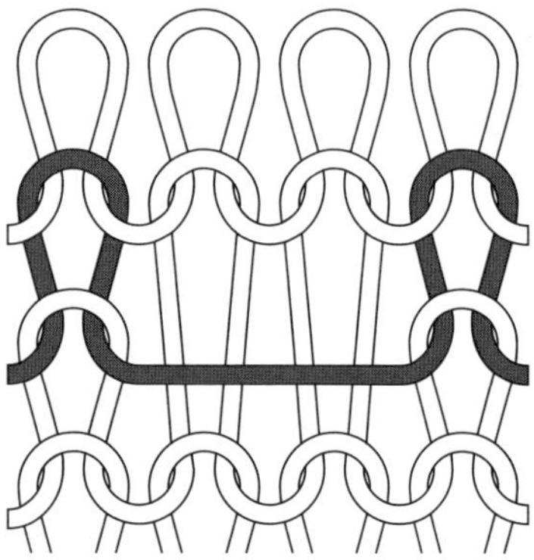

Figure 12.4 Miss stitch

Classifications of Knit Fabrics

Knit fabrics are classified into two basic types according to their construction processes. These two broad categories are known as weft knits and warp knits. Each of these types of knit is manufactured using different equipment, usually in different factories and for different industries.

Weft or filling knits are made in the weft, or horizontal, direction. Because they are made horizontally, they will unravel only in that direction. Hand knitting is weft

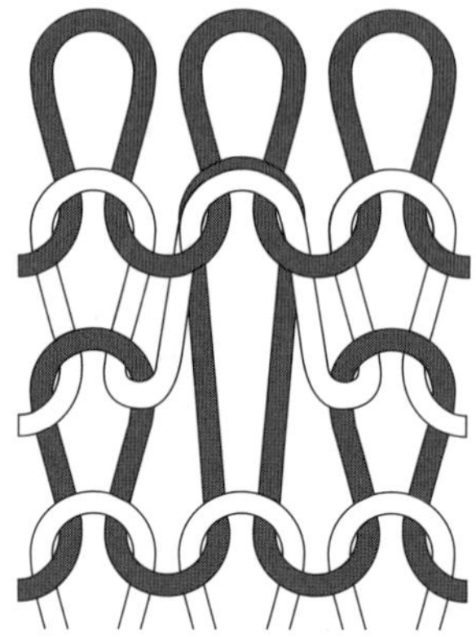

Figure 12.5 Tuck stitch

knitting. This is still the most common type of knitting, although warp knits are increasing in number. Both warp and weft knits use only one set of yarns, but weft knitting allows an entire fabric or product to be made from a single yarn. Weft knitting is most often used in the sweater, blanket, and sheeting markets.

Warp knitting is constructed in the warp, or vertical, direction. Each wale requires its own yarn. Although the yarns run vertically, they do so in a zigzag down the fabric. The fabric will not run or ravel due to this construction. Warp knits are most appropriate for window and table laces, and tricots. This is a fast-growing market today. Warp knitting is more thoroughly addressed in Chapter 13.

Figure 12.6 and Table 12.2 show the specific differences between the weft and warp knits.

Basic Weft-knit Fabrics

Jersey is the simplest of the weft knits, made of all knit stitches. As such, it has wales on the front and courses on the back. Notice that since it is made of all knit stitches, the fabric is not stable and curls to the face of the fabric. (Have you ever cut a T-shirt?) Jersey stretches pretty evenly in both directions. Shrinkage can be a problem. This fabric is used primarily for soft, cozy sheets.

[Reference Swatch 98]

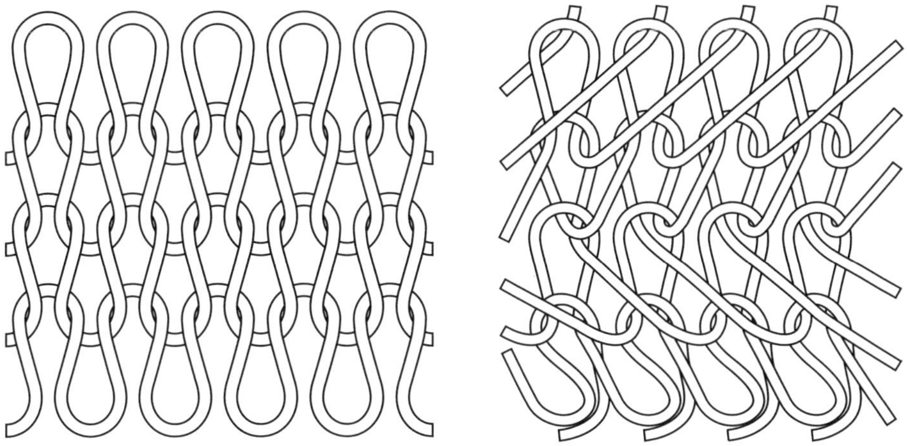

Figure 12.6 Comparison of weft and warp knits

Table 12.2 Comparison of Weft and Warp Knits

Weft (Filling) Knits	Warp Knits
Loops and fabric formed horizontally	Loops and fabric formed vertically
Each needle makes one loop at a time	Entire row of needles makes all loops at once
Connections are horizontal	Connections are diagonal
One continuous yarn	Series of yarns
Flat or circular machines	Flat machines only
Finished edges (selvages) or tubular form	Finished edges (selvages) only
Hand or machine process	Machine process only
Tendency to run in the wale direction	Does not run
Unravel horizontally	Impossible to unravel
Each needle can be individually controlled	Higher productivity—faster machines
More structural designs possible	Structural designs are possible with raschel machines only
Two-way stretch possible	Crosswise stretch, little lengthwise stretch (can be produced with no stretch)

Rib Knits

Rib knits are made of knit and purl stitches. Wales are evident on both sides, and the courses are harder to see. These fabrics are more stable and do not curl. They have the best elasticity (which means recovery) in the width direction. Waffle knits are a rib variation that provides superior, lightweight insulation that is an excellent choice for blankets.

Double Knits

Double knits are the most stable of knits; they are essentially two fabrics knitted together. This makes them good choices for small upholstered pieces, accent pillows, and blankets, as they can be completely reversible. Mattress pads are often double knits.

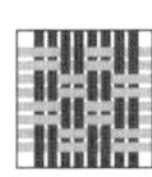

[Reference Swatch 99]

Pile Knits

Pile knits are the second of the pile structures, and just like pile weaves, they use an extra set of yarns to create loops that form a three-dimensional surface on the fabric. These loops can be preserved, or sheared and napped (brushed), for a more luxurious, softer surface. These fabrics are chosen for blankets, throws, and accessories.

[Reference Swatchs 100–102]

Study Guide

Table 12.3 offers a quick reference guide to assist in identifying weft knits. You will find Table 12.3, Quick Reference Guide for Weft Knits, preceding the swatch boards. Please remove the table and place it within Chapter 13 for easy referencing.

Activity 12.1 Knit Fabric Analysis

Name: ______________________________ Date: __________

Objective: To identify the components of a knit fabric and its properties.

Materials required: Knit swatch provided by your instructor.

Procedure

1. Mount the swatch provided by your instructor, face up, with the warp placed vertically, attached only at the top.
2. Using your pick glass, knowledge from the burn test in Activity 4.1, and weight computations in Activity 6.1, analyze the swatch.
3. Record the following information:
   a. Fiber content: ______________________________
   (e.g., 100 percent cotton)

   Yarn construction: ______________________________

   (Spun or filament? Single or plied? If plied,how many?)

   Fabric name: ______________________________
   (e.g., jersey, sliver knit, double knit)

   Fabric count: ______________________________
   (e.g., 88x76)

   Weight: ______________________________
   (expressed in oz. per sq. yd.)

Activity 12.2 Fabrication Worksheet

Name: ______________________________ Date: ____________

Instructions: Fill in all fields of the following chart.

The black fields do not need to be filled in. (Count only 19 and 127.)

Swatch #	Woven or knitted?	Fabric structure	Fabric name (if different from structure)	Count	*Possible end use of fabric
6					
14					
19					
20					
22					
30					
122					
127					
134					

*Provide a more specific "use" than is provided in the book.

CHAPTER THIRTEEN

Warp Knits

Reference Swatches

103. Tricot | 104. Brushed Tricot | 105. Sueded Tricot
106. Minky | 107. Tulle | 108. Raschel Lace
109. Raschel Crochet | 110. Point d'Esprit

Warp knits differ from weft knits in their method of construction. Warp knits get their name because they are made in the warp, or vertical, direction. Using a whole series of yarns, these yarns are fed to the needles vertically. As the needle beds shift from side to side, the yarns zigzag down the fabric. Because of the zigzag construction, warp knits cannot be made by hand, and they will not run or ravel.

Warp knits are generally thinner and lighter weight than their weft-knitted counterparts. They are smoother, more shrink resistant, and less resilient. They are also a very stable knit (see Table 13.1).

Warp knits are divided into two major categories: tricot and raschel (see Table 13.2).

Tricot

Tricot is the simplest of the warp knits. Notice that Swatch 103 has a different face and back, just like jersey. In fact, tricot and jersey are made quite similarly. Each is made entirely of knit stitches, with a face of all wales and a back of all courses. But jersey and tricot are made on different machines and therefore do not look the same. Generally, tricot is lighter weight and smoother than jersey because it is usually made of filament yarns. Unlike jersey, tricot does not run or ravel. Note the loops, or Vs, on the front, and the laps, or diagonals, on the back. Sometimes, this looks like a vertical rib on the front and a horizontal rib on the back. Compare this tricot to a simple jersey, Swatch 98. Tricots are often used as backings for bonded fabrics, such as pleathers, imitation suedes, and even wallcoverings. Sueded and brushed tricots can be excellent choices for upholstery, as there is little stretch, great recovery, and they do not run. Since they are primarily made of synthetics, they are also easy to care for (see Figure 13.1).

[Reference Swatches 103–106]

Raschel

Raschel is the most complex of knit structures, often called meshes, nets, or laces. This fabric is a knitted alternative to lace structures and is the most common lace-type fabric

Table 13.1 Contrast of Warp and Weft Knits

	Warp Knits	Weft Knits
Stitch direction	Vertical	Horizontal
Stability	Does not run or ravel	Runs in the wale; unravels horizontally
Yarns	One yarn for each wale	One yarn for whole fabric
Stretch direction	Mostly vertical (unless cut cross grain); spandex creates four-way stretch	Mostly horizontal, but both directions have some stretch
Machinery	Flatbed only—faster technology	Circular or flatbed

Table 13.2 Comparison of Tricot and Raschel Knits

	Tricot	Raschel
Machinery	Basic construction High speed Fine gauge	Complex construction Fine to coarse gauge
Yarns utilized	Fine denier filament and textured filament	Spun and filament All counts and deniers
Design potential	Simple designs No holes or lacy effects in fabric (except athletic mesh)	Large and intricate designs Lacy effects in fabric
Typical fabric end uses	Lightweight bedding, blankets, and upholstery Home decoration for novelty furs	Laces and lace trim, nets, tableware, window treatments, blankets, carpeting

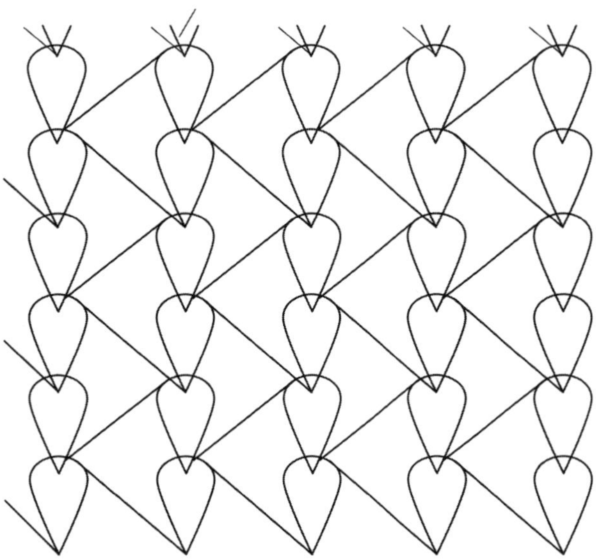

Figure 13.1 Warp knit structure—notice the zigzags between stitches

used for casement type window treatments. Most of what we call lace is really a raschel knit.

Although raschels are most often made of filament yarns, there are also spun variations, such as an imitation crochet, made on a raschel knitting machine. A filament yarn aids in providing strength to a rather ephemeral lacey construction, and a textured filament could be used to help lock the stitches together for even greater durability.

Raschels can also be used for thermal blankets and mesh laundry bags.

See Table 13.2 for a comparison of warp knits and their appropriate uses.

[Reference Swatches 107–110]

Study Guide

Table 13.3 offers a quick reference guide to assist in identifying warp knits. You will find Table 13.3, Quick Reference Guide for Warp Knits, preceding the swatch boards. Please remove the table and place it within Chapter 13 for easy referencing.

Activity 13.1 Fabric Structures: Knits

Name: ______________________________ Date: __________

Attach a 3 × 4-inch swatch of each knit construction listed below, technical face up, with the warp in the vertical direction. Do not use swatch book swatches. Do not share swatches with another student.

Jersey Knit	Rib Knit—Fabric Name: __________
Sueded or Brushed Tricot	Pile Knit—Fabric Name: __________
Tricot	Raschel

CHAPTER FOURTEEN

Minor Fabrications

Reference Swatches

111. Imitation Suede | 112. Needlepunched Felt | 113. Pleather
114. Oilcloth | 115. Quilted | 116. Embroidered Eyelet
117. Tufted Chenille | 118. Venise Lace | 119. Leather

Minor fabrications is a collection of several different construction methods. Effectively, it is the miscellaneous category of fabrics, and it includes all other methods of construction that are not woven or knit. Although these fabrics are called minor, do not underestimate their significance: Their manufacture adds up to billions of yards per year. These fabrics are the bones of the interior design industry and include such hidden structural fabrics as carpets and underpadding, as well as trims, laces, embroideries, and films. The fabrics in this chapter are organized by construction category.

Fabrics Made without Yarn

The fabrics in this category are made without yarn. Without yarns, there is no organized structure and no thread count. These fabrics are often softer and more elastic than their woven counterparts. As yarn making is the most time consuming process in the production of fabric, fabrics made without yarns are also less expensive to produce.

There are several kinds of fabrics made without yarn. Many imitation suedes are nonwoven, made from an entanglement process (Swatch 111). Notice that there is little stretch and no grainline. The most common trade name for imitation suedes is Ultrasuede®. This can be quite expensive, as it is created by a foam type of extrusion process. Other suede types are woven or knitted first and then emerized or sueded as a finish. They have a single sueded side. Ultrasuede® or imitation suede is suede-like on both sides.

Imitation suede is more durable than actual suede. As it is made from polyester, it can be thrown in the washer or easily cleaned. It is also stronger, dimensionally stable, and more resistant to cat scratches.

Nonwovens are prevalent in the disposable market, such as dryer sheets, cosmetic pads, disposable diapers, filters, kitchen wipes, and hospital gowns and masks. Disposables are preferred when the cost of sterilizing is prohibitive.

Durable nonwovens include indoor/outdoor carpeting, carpet underlay, and battings. Battings are the stuffing of pillows, quilts, and upholstery. It is important that these structures are hydrophobic (stain resistant), with good compressional resilience, and that they do not sag or become lumpy over time. It is also important that they have good loft and easily recover from compression.

Carpet underlay is typically a nonwoven. Making the right choice is important. The underlay is the support for your carpeting and assists in determining the longevity of the carpet. As a general rule, simple foam slabs are cheap but ineffective. Memory foam pads, however, can be a good choice, as they are engineered to continuously recover. Foam slabs, including memory foams, are produced by introducing bubbles in a viscose solution. This process gives the foam dimensional stability over a long life.

Carpet underpads are often sold by weight density. A higher density is generally stiffer but will last longer. The thickness of a pad relates to the carpet choice: thinner is better for a Berber; thicker is better for a plush. Today there are new finishes, such as antimicrobial, stain, and static resistance, which should also be investigated.

Felt and felted fabrics are also made without yarns. Felts are made by applying heat, moisture, and agitation to a minimum of 50 percent of animal hair. The fibers are matted together by this process. Swatch 112 is made of polyester (no animal hair) and so is a nonwoven, made using the entanglement process.

Thick felts are used for carpet underlay, insulation, and upholstery, depending on the fiber content and quality of the finished goods. Thinner felts can be used for wallcoverings, pillows, and table covering, as well as some drapery. Since felt has no grain, it can be formed into a three-dimensional shape for pillows or accessories. Felt does not fray or ravel and has excellent shock and sound absorbing properties.

Pleather is usually a bonded fabric, the backing is most often a nonwoven or a tricot. The fabric is subsequently embossed for surface interest and provided with a shiny colorant. This fabric is designed for easy clean-up.

[Reference Swatches 111–113]

Fabric Combinations

Fabrics in this category are made with simple knit or woven fabrics that are sewn or fused together to form new fabric combinations. These combinations have weights, values, and end uses different from the individual fabrics themselves. Oilcloth and pleathers are often bonded. The bonding or backing adds strength, stability, and longevity. Backings are often nonwovens or tricots, although many other backing solutions are possible (Swatches 17 and 102).

Swatch 115 is a quilted fabric consisting of three layers of fabric held together by a decorative stitch. Quilted fabrics are used for bedding, upholstery, and accessories. The face and back can be almost any fabric; the center layer is a nonwoven batting.

[Reference Swatch 105]

Fabrics Made with Yarn

Fabrics in this category are made with yarn, but not woven or knitted, and include embroidery, tufting, and lace structures.

Embroidery is a surface embellishment applied to a woven or knit fabric. This embellishment can include embroidery stitches, beading, sequins, feathers, studs, and more. Sometimes the fabric is further embellished with eyelets or holes (Swatch 116). The fabric could also be manipulated to create dimension as in the pintucked example in Swatch 26.

Tufting is the process of inserting yarns into an existing grid, creating a three-dimensional structure. Carpeting is often made this way, as is the tufted chenille in this chapter (Swatch 117).

Original laces were developed in convents in the middle ages and often take their names from these convents. This includes laces such as Nottingham, Cluny, Chantilly, and Alençon. Lacemaking is a slow process made by knotting, looping, or twisting yarns. Lace is the most complicated of all textile processes and therefore the most expensive of all fabrics. The example provided in this text is Venise (Swatch 118).

One of the most important considerations when purchasing lace is how to order it. Lace can be a tiny trim or full-width yardage; it can have straight or scalloped edges. A "putup" is an industry term that relates to how the fabric is packaged: the size and shape in which the fabric is sold. Putups are always a consideration in fabric acquisition, but even more so for lace, as it has so many permutations and possibilities. One could purchase a raschel flounce or an Alençon galloon, for example. The raschel and Alençon are the lace types, according to construction; the putups are the packaging, such as the flounce or galloon in the preceding example. Refer to Table 14.1 for an illustration of putups for lace.

[Reference Swatches 116–118]

Table 14.1 Lace Putups

Putup Name		Description	Width
Allover lace (tablecloth lace)		Full-width lace yardage; used for curtains, accessories, and tablecloths	36 to 72 inches
Flounce		One single decorative or scalloped edge	½ to 18 inches
Galloon		Two opposing, or mirror-image, decorative or scalloped edges	Varies
Swing	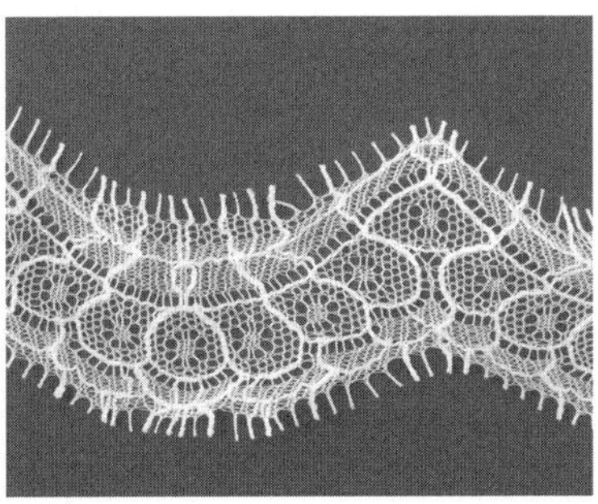	Two nesting decorative or scalloped edges	Varies
Medallion		A single motif (like a doily or a heart-shaped piece of lace)	Varies (not yardage)

Fabrics Made without Yarn or Fiber

Fabrics in this category are made directly from the liquid solution. The solution is poured onto a collection surface, creating a flexible sheet, much like a shower curtain. Fibers and yarns are not used. Films, pleathers, and oilcloths are some of the fabrics in this category used to make shower curtains and upholstery. They are often called vinyl, which is, in fact, a fiber just like polyester and not a fabric name. Film is a thin sheet of impervious acrylic, vinyl, PVC, or polyester. The oilcloth presented in this text does have a grid (made with yarns) embedded into the film for durability, as thinner films can be quite flimsy (Swatch 114).

Pleather is a bonded film that is an imitation or synthetic leather. In the past, there were many negative connotations associated with it, such as plastic, cheap, and environmentally unfriendly. Rebranded as vegan leather, pleather is now desirable on a number of fronts. First, no animals are harmed in its manufacture. Pleather is perceived as being cruelty free, comfortable, convenient, economical, and even trendy. Additionally, it is so much easier to care for than actual leather. Spills can just be sponged off.

Skins

Leather and suede are skins from an animal. Today we use the skin of cattle (primarily), snakes, pigs, lambs, alligators, ostriches, and many other animals. Fish skins are even beginning to find a market. The tanning processes of leather making (preserving and converting rawhide to leather) are generally harmful to the environment. Sustainable alternatives, such as vegetable tanning (as opposed to chromium tanning), are being explored for this reason. Due to environmental concerns and potential issues of animal cruelty, consumers are sometimes pausing before using actual leather.

Leather is essentially the outside of the skin; suede is the inside. As such, leather has a smoother surface and is easier to care for. Leather gets more beautiful with age, but does need to be oiled regularly to retain its supple hand and extend its life span. As commercial applications do not use oiled leathers, new treatments and finishes have been developed, including crypton, water repellent, and antimicrobial. Figure 14.1 illustrates leather quality according to its location on the animal. (Swatch 119).

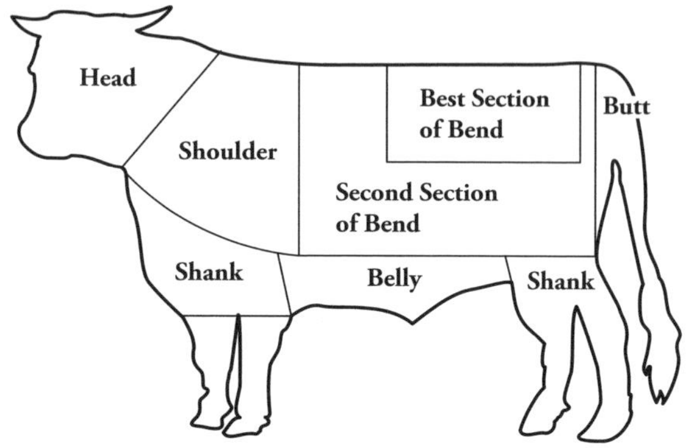

Figure 14.1 Leather quality

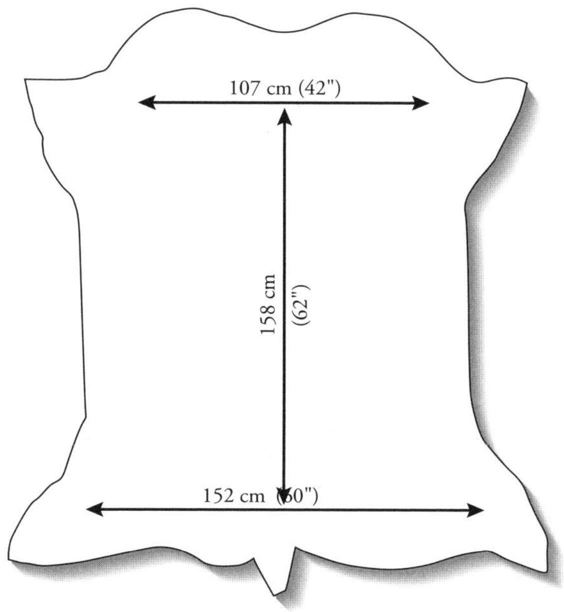

Figure 14.2 Hide yardage calculations

Purchasing leather is a bit complicated, as it is a hide and not yardage. The hide is laid out, and measurements are taken from the widest parts. Square footage is calculated from these areas, and the price is determined by this square footage. Note that although you will pay for this larger number, the hide does not, in fact, yield the number of square feet for which you will be asked to pay (Figure 14.2).

There are four grades of leather:

Full Grain comes from the top layer of the hide, which has all of the grain. Therefore it is called full grain. Full grain leather burnishes and beautifies with use. Best quality.

Top Grain is the second highest grade. It is essentially the second layer of the skin. Top grain has been split

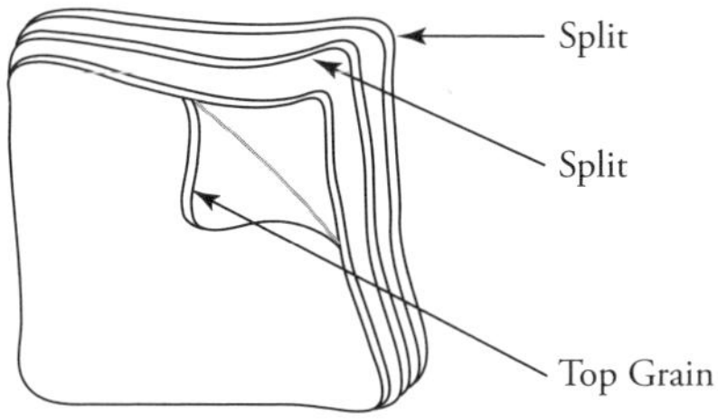

Figure 14.3 Leather splits

from the first layer of blemished hide and has been sanded to a beautiful finish. This is how scars, brands, scratches, or injuries are eliminated from the surface of the fabric. For this reason, it is sometimes called corrected grain. Strong, but does not age as well as full grain. Figure 14.3 illustrates splits.

Genuine is the third layer and is generally called suede. Suede is produced from leftover layers of the hide, once the better layers are split off. The surface is sanded to a softer, more velvety, hand and appearance. Suede is not as strong and does not respond well to liquids.

Nubuck is a leather that has been sanded to make it suede-like. As it is a leather, it is stronger and more durable.

Bonded is the fourth grade of leather, made from leftover scraps that are bonded together with glue, and resurfaced in a process similar to vinyl manufacturing. Most Bibles are covered with bonded leathers.

Suede is often being replaced by Ultrasuede or sueded finishes on wovens and knits. Imitation suedes are often better for the environment, easier to care for, and offer better color choices.

For a summary of minor fabrications, see Table 14.2.

[Reference Swatches 111–119]

Interior Design Focus: Minor Fabrications

Window Treatments

Laces and embroidery are excellent choices for windows (both drapes and sheers) and tableware, as well as for trims and embellishments throughout the home.

Sheets/Bedding

Several of the fabrics presented in this chapter are good choices for the bedroom. Tufted chenilles, and the contemporary variation called Minky, are often found as bed toppers. Pintucking, embroidery, quilts, and mattress pads are also widely used in this category.

Table 14.2 Minor Fabrications

Type	Construction	Examples and Fabrication Methods
Made without yarn	Made directly from fiber; no yarn involved	Nonwovens: Made by adhesion, heat fusion, or entanglement, Ultrasuede Felt: Heat, moisture, and agitation applied to at least 50 percent animal hair
Fabric combination	Made by two or more fabrics fused, stitched, or adhered together as one	Bonded/laminated: Pleather, oilcloth, many bonded upholstery fabrics Quilt: Three layers held together by a decorative stitch
Made with yarn	Not made by weaving or knitting	Embroidery: Surface embellishment. Decorative stitch applied to the surface of a fabric. Includes beading, sequins, and other stitched fabric manipulations Tufting: Inserting yarns into an existing grid; used in carpeting and tufted chenille Lace: Made by looping, twisting, or knotting; types include Leavers, Nottingham, Cluny, Alençon, and Chantilly, among others
Made without yarn or fiber	Made directly from solution, nonfibrous	Films like pleather, Gore-Tex®, Neoprene®
Skins	Harvested from animals	Cow, pig, lamb, ostrich, alligator, many others

Upholstery

Ultrasuedes, felts, pleathers, quilts, leathers, and films are excellent choices for upholstered pieces. These are heavier weight structures, which support the performance and durability requirements of upholstery applications.

A Brief Carpeting Primer

Carpeting requires separate study; many detailed texts are devoted to this subject. As it is a textile, a brief overview is in order.

Carpets can be woven, knitted, or tufted. Essentially, it is a carpet if secured to a floor and a rug if it is loosely laid with bound edges. There are many areas of consideration when placing the right carpet in a room. The first of which are the inherent traffic patterns expected of the room. A softer, but less durable, carpet can be used in a spare bedroom or as a scatter rug. When durability and high traffic are expected, such as a staircase or family room, both fiber and carpet construction must be closely scrutinized. Carpet rating systems (which rate carpets between 1 and 5) help determine the best carpet for the intended use. A rating of 5 is the most suitable for high traffic.

The primary carpet fibers are nylon and olefin but others include polyester, cotton, wool, rayon, and acrylic. Ninety-nine percent of all carpet fibers are synthetic. Fiber and fabric should be carefully considered for each application, as some fibers crush more than others (poor resilience), some absorb oil stains more than others (oilophilic), and some should be used in places where there is low traffic only. These fibers and their individual benefits are discussed in Table 14.3.

Table 14.3 Carpeting Fibers

Nylon	70% of carpet market Excellent abrasion resistance Can be treated to be soil resistant Can be treated to be antimicrobial Can be treated to resist pet stains Electrical conductor Dyes well Colorfast to cleaning Colorfast to sunlight Excellent crush resistance Mildew resistant
Olefin	12% of carpet market Low cost Inherently stain resistant Excellent abrasion resistance Colorfast to sunlight Best choice for indoor/outdoor carpets Solution dyed only Electrical conductor (better than nylon) Crush resistance (less than nylon) Chemical, moisture, & stain resistant Mildew resistant
Polyester	13% of carpet market Inherently stain resistant Excellent colorfastness Resistant to water-based stains Luxurious hand Less durable than nylon Continuous filament fiber makes it more durable Good for low traffic residential—prone to crushing Bleach resistant UV resistant Mildew resistant Can be recycled
Acrylic	Very woollike Excellent bulk and cover Residential only Soft and lightweight Crushes easily Less abrasion resistant Less resilient Oilophilic Mildew resistant Best used for bath mats and scatter rugs
Cotton	Soft hand Expensive Mildews Best used for bath mats and scatter rugs
Rayon	Soft hand Shiny Mildews Best used for bath mats and scatter rugs
Wool	1% of all carpets but all other fibers emulate it Expensive Excellent insulator Good abrasion resistance (compromised with small fibers like Merino) Luxurious hand Resilient Reduced soil Self-extinguishes—chars but does not burn Ages well

Yarn twist should also be evaluated. As in chapter 5, the higher the twist per inch, (TPI), the more durable the carpet, and the longer it will look new. If the yarns untwist easily, the yarns will lay down, and the carpet will begin to look worn. This will be one of the reasons that two carpets that look similar will have different price-points. Hard-twisted yarns are evidence of quality and longevity in a carpet.

A bulked continuous filament (BCF) yarn is a texturized yarn that increases bulk, loft, and texture in the ultimate carpet.

Thread count, or density (the number of threads or knots per inch), will also have a huge impact on performance. The higher the count, the closer together the yarns are, which leaves less space for the fibers or yarns to lay down or mat down. A tightly woven carpet will stay fresh looking longer. Higher density carpets may not feel as soft to the hand but will generally retain their appearance longer.

Carpet structures will also impact the performance, placement, and ultimate purchasing decisions. See Table 14.4 for a discussion of carpet types.

Table 14.4 Carpet Structures	
Plush Textured Carpet 	Cut & uncut pile Stiffer hand Does not show footprints or vacuum marks Will keep its appearance longer Textured
Plush Saxony Carpet 	Cut pile Very soft hand Has some luster Yarns are the same height Yarns are well defined due to twist
Plush/Velvet 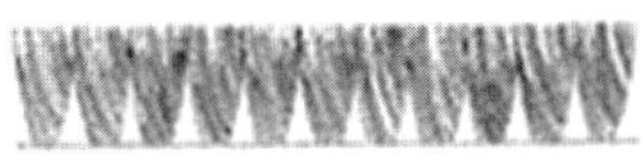	Shorter cut pile Smooth surface Shows footprints, vacuum marks Low-twist yarns fluff out
Berber Carpet 	Woollike look Yarns are looped, loops are not cut Very durable to traffic Stiffer, rougher hand (or foot) Can run (if pets scratch at it)
Patterned Carpet	Can be plush or Berber with a pattern
Printed Carpet	Printed plush carpet
Level Cut Loop 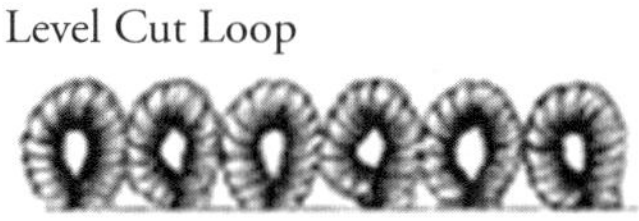	Cut & uncut loops to create a pattern or texture Multilevel loop also possible
Frieze 	Cut pile Tightly twisted yarns that curl into each other in a random way Textured, uneven surface Reduces footprints Hides soil well Easy to maintain, great durability and resilience
Shag 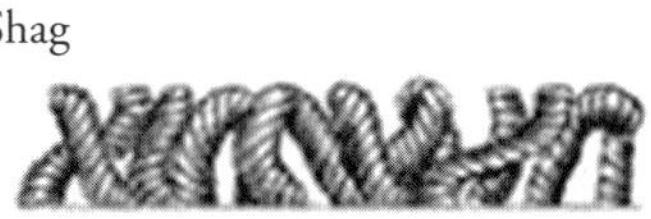	Cut pile Loosest carpet structure—resembles a modern shag carpet—low density Long, floppy yarns—deep pile, but they do not stand up More difficult to clean
Woven carpets: Wilton, Axminster	More expensive More durable Better patterns, lots of color

Activity 14.1 Carpet Analysis

Name: ______________________________ Date: ____________

Extend your carpet knowledge by shopping for carpet. Go to at least two carpet outlets to complete your research. Find two remarkably different carpet choices for a residential placement and two for contract.

Analyze the details listed in the chart below with as much information as you can acquire at the store and on the labeling. Attempt to get samples for verification and to attach to your report.

Prepare a report with your recommendations for each. Support your conclusions.

	Residential carpet #1	Residential carpet #2	Contract carpet #1	Contract carpet #2
Fiber content				
Type of carpet— pile, flat, loop, Berber, frieze, Saxony				
Yarn type				
Pile density/count				
Backing materials				
Backing structure				
Finishes—stain, crush resistance, etc.				
Warranty				
Price per yard				

CHAPTER FIFTEEN

Dyed and Printed Fabrics

Reference Swatches

120. Fiber/Stock Dyed | 121. Yarn Dyed | 122. Piece Dyed
123. Calico/Ditsy | 124. Paisley | 125. Toile de Jouy
126. Direct Print | 127. Blotch Print | 128. Overprint
129. Discharge Print | 130. Flock Print | 131. Flocked Dotted Swiss
132. Velvet Burnout | 133. Flat Burnout | 134. Duplex Print
135. Tie-dye | 136. Batik | 137. Ikat

Color is the most important factor in marketing textiles. Consumers are often more concerned with choosing the correct color than with durability and performance. Color and print designs are additional factors in fabric identification. Print orientation, type of print, and stage of dyeing will impact consumer selection and therefore must be considered as fabric evaluation criteria.

The Basic Dye Process

Most of us have some understanding of the dye process, whether it is from dyeing our hair or even staining a favorite dress. The process is simple: Powdered dyestuff is dissolved in a small amount of hot water. This creates a stock solution, which is highly concentrated liquid dye. The stock solution goes into a pot along with an activator, which is a group of chemicals that makes the color brighter, permanent, and consistent. Add this to the correct amount of water to suspend the dye solution. The more absorbent the fiber, the better it will absorb the water and, along with it, the dye. Extending the dye time results in better dye penetration but does not really affect the final color. The ultimate color is decided while mixing the ratio of dye to water and the ratio of dye to fabric, called the liquor ratio.

The fabric and dye mixture are brought together at fairly high temperatures for a few hours to complete the dye process. Dyers are often chemists, who can calculate the correct color for any fiber. Dyers work to make permanent chemical bonds between the fiber and the dye, and the chemicals required to bond dye to silk are different from the chemicals required to bond colorants to polyester. Each fiber has a specific affinity to a particular "dye class." There is no single dye class that can dye all fibers.

There are many regulations about the dyeing process today, particularly about minimizing waste and the disposal of toxic pollutants. For this reason, dyeing is often a separate area of specialization. Fabrics are often sent to a converter for color or printing after the textile manufacturing process. A converter is a specialist that dyes, prints, or finishes the fabric. Large, vertically integrated companies may have an in-house dyeing department.

Dyes that use less water and energy are being developed and incorporated in the coloring process in an attempt to reduce the environmental impact of coloring fibers and fabrics. These include waterless dyes, non-VOC (volatile organic compound) dyestuffs, carbon dioxide dyes, and low-impact fiber-reactive dyes.

Stages of Dyeing

Color can be applied at any stage in the life of a textile: there are four different distinct opportunities for the application of dye in production. The dyeing can be accomplished while the fabric is in the fiber stage, the yarns can be dyed, or it can wait until the fabric or even the garment stage. The latter, of course, belongs only to the apparel industry. Each of these methods creates a different color effect along with different levels of thoroughness and expense.

Locate the denim swatch (Swatch 56). Pull a blue yarn out of this swatch and untwist it; you should see white in the center. This is because the yarn was made without color. The yarn was subsequently dyed, and because the yarns were tightly twisted into a yarn, the dye did not get into the center of the yarn. This is an example of poor dye penetration, and it sometimes occurs when dyeing is accomplished later in the production cycle. The result could be improved by dyeing the yarn longer or by dyeing the fiber before we made the yarn.

Fiber or Stock Dyeing

The earliest possible moment for a natural fiber to be dyed is in the fiber stage, often called stock dyeing. The loose fibers are dyed, and the yarn is then constructed from different colored fibers. Tweeds and heathers are dyed in this way. Notice the variations in color on the fabric and ultimately along the length of each yarn. This process is the most thorough method of dyeing; the color is fully penetrated. It is also the most expensive stage of dyeing, due to loss of fibers in the process. Finally, color choices are made early in the production cycle and may not represent the current trends (Swatches 35 and 120).

[Reference Swatch 120]

Solution or Dope Dyeing

For a purely synthetic fiber, the earliest stage of dyeing is in the solution stage, while the fiber is still liquid. The liquid is called the solution, or dope. Dope dyeing is the most thorough and the most colorfast (retains color) method for synthetics. Once again, dyeing at this stage is expensive. The final appearance is a solid color. Solution dyed products are colorfast to bleach, so this is a good alternative for carpets and areas where bleaching is necessary.

Yarn Dyeing

The yarn stage is the second possible stage in which to dye for all types of fiber. See Figure 15.1. At this stage the fiber has been made into yarn, and it is the yarn that is being dyed. As we saw with the denim example, this method is not as thorough as dope or fiber dyeing because it takes longer for the dye to penetrate to the center of the yarn. Yarn dyeing is still a good quality stage of dyeing, producing beautiful fabrics that will endure longer than prints. This is the preferred dye method if you are looking to create fabrics that are striped, plaid, or checked, as in gingham (Swatch 42) or chambray (Swatch 43). Yarn dyeing is also the method used for tapestries and other jacquards (Swatches 79, 82–85). Dyeing at this stage produces better quality than the next stage of dyeing.

[Reference Swatch 121]

Piece Dyeing

Dyeing the whole fabric when it is made of a single fiber is called *piece*, or *fabric*, dyeing. Piece dyeing is less thorough and less expensive than the previous methods of dye. The whole bolt of fabric is being dyed, and the visual result is a solid-colored fabric. Certainly solid-colored fabric could have been made with colored yarn, and therefore yarn dyed. The fabric could also have been dope dyed, but these are more expensive alternatives. When evaluating fabrics and determining the dyeing stage, assume that the manufacturer will choose the most economical process for the desired result. Solid-colored fabrics are most often piece dyed, and piece dyeing is the most common stage of dyeing.

See Figure 15.1 for a graphic representation of the stages of dyeing.

[Reference Swatches 120–122]

Basic Dye Chemistry

No single dye class is capable of dyeing all fibers. This means that one must understand simple chemistry in order to determine the correct dye for each fiber. Once the chemical base of the fiber is ascertained, the appropriate dye class can be selected. For example, cellulose fibers generally dye best in a fiber-reactive dye class, whereas protein fibers are best dyed in an acid-dye bath. Neither of these dye classes work with polyester.

Another consideration is the choice of dye versus pigment. These are both colorants with some subtle but relevant differences, as evidenced in Table 15.1. Making the correct choices in terms of color media is as big a decision as any of the other elements of a textile.

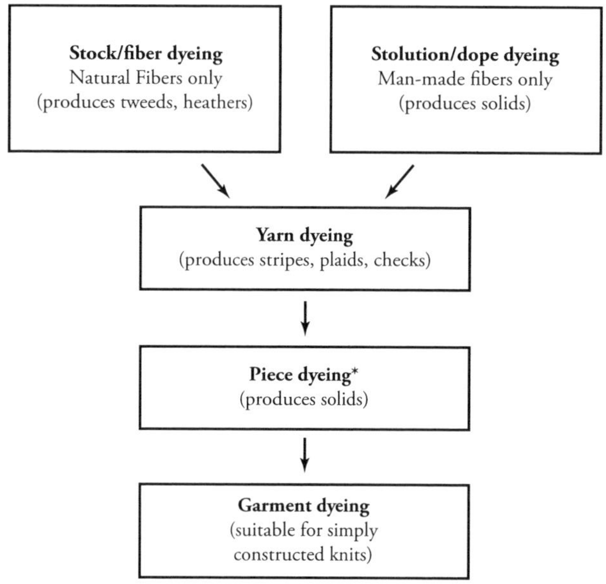

Figure 15.1 Stages of dyeing

Table 15.1 Properties of Dyes and Pigments

Dyes	Pigments
Transparent	Opaque
Aqueous (water soluble)	Nonaqueous
Immersion	Topical (surface)
Soft hand	Stiff hand
White not available	White is available
Colorfast to laundering	Colorfast to sunlight and bleach
Requires dye affinity	No affinity required
Requires dye-lot matching	Excellent lot-to-lot matching
Aged (finished with heat, steam, and water)	Cured (finished with dry heat)

Special Dye Processes

We have determined that there is no single dye capable of dyeing all fibers and that we need precisely the correct dyestuff for each dye project. When different fibers come together in a blend, a more creative solution must be sought. A poly/cotton tablecloth, for example, would require two different dye classes to dye the tablecloth a solid red. This treatment is called union dyeing. To accomplish this, two classes of dyes (one for cellulose and one for polyester) in the same color would be put into the dye pot, along with any necessary chemicals or activators. The tablecloth (or a group of them) would be placed in the dye pot; the cellulose dye would dye the cotton, the synthetic dye would dye the polyester, and the tablecloth would come out a solid red. In short, two or more fibers plus two or more dyes equals one color (2 + 2 = 1).

Using the same technology, we can dye a fabric made of two (or more) fibers, two (or more) different colors, in the same dye pot, at the same time. Two different dye classes in two different colors are placed in the dye bath, and produce a fabric made of two (or more) different colors.

The outcome appears yarn dyed, yet it was dyed in a piece-dyeing stage. Bolts of fabric can be kept in inventory as a white fabric, called PFD (prepared for dye). This means that the fabric is white and clean of any finishes that might inhibit the dye process. The PFD fabric is dyed at the last moment (before sales) in several different color combinations. In this way, the colors are kept relevant and current. This effect is called cross dyeing. In this case, two (or more) fibers plus two (or more) dye classes equals two (or more) colors (2 + 2 = 2).

Color Management

Color management plays a huge role in the marketplace today and includes a concern for color matching and color longevity. Today's standards are such that the colors not only have to match, but they have to stay matched! The following are some relevant industry terms pertaining to color management.

- **Dye lot** refers to a specific, unique dye bath. Everything that shares the dye lot's number is dyed at the same time, with the same recipe and the same bath. This provides assurance that the colors are perfectly matched. When developing a collection, it is imperative that all the pieces in the collection or, more important, from the same sofa, come from the same dye lot.

  A sofa might require 20 yards of fabric. As the fabric is expensive, it would be tempting to use the last 15 yards of one bolt before opening the next. Although the two fabrics may appear to match under the showroom lights, they may look quite different when placed in the home.

- **Greige goods** (pronounced gray) is unfinished fabric as it comes off the loom or knitting machine. Most fabric is not created in color. Nor are most fabrics white in the beginning. Muslin is the color of greige goods, and most fabrics begin this way. Bleach is added later to create white fabric, or fabric is sent to a converter to add color.

- A **lab dip** is a swatch dyed in a lab to meet a color standard provided by the client. The lab dip would be created in a specific fiber and fabric type, for client approval. Several lab dips could be attempted before successfully achieving the standard and final approval of the dip. These approved dips are carefully controlled and preserved and would be referred to every time an order is placed by the customer.

- A **spectrophotometer** is a computer that matches color. This device will scan an existing color and provide a recipe for it in any dye class desired. A spectrophotometer will also verify that one color matches another by digitizing the two colors, giving a degree of certainty greater than the human eye can see. A spectrophotometer is more reliable than previous methods of color matching that rely on visual inspection of fabrics, such as a fan deck, or colors on a computer screen that may not have been calibrated in a long time.

- **Metamerism** is a phenomenon of light whereby colors appear different under different light sources. Consider the case of coordinating separates, all black. In the morning, all the pieces look the same, but once in the sun, two of them no longer appear to match. In this case, the two blacks appeared the same under the incandescent lightbulb that is in your bathroom but did not match when viewed in direct sunlight—an interesting concept, considering that we shop, go to school, and work under cool, white fluorescent bulbs, live under incandescent lightbulbs, and crave the yellow of direct sunlight.

  If we apply this concept to coordinates in a room, the results could be disastrous. If the carpeting is a significantly different red than is found in the print of the sofa, the client will not be happy.

- A **Macbeth light** is a light box that has several bulbs providing all the light sources in it. This is used to evaluate the colors that are meant to live together and to detect potential metamerism. Often, big-box home-improvement stores offer a small light box to illustrate significantly different colors of light.

- **Colorfastness** is the ability of a fabric to resist color loss. A variety of factors can contribute to color loss, including laundering, bleach, perspiration, swimming pool chlorine, fume fading (environmental gases), sunlight, and frosting (localized abrasion). A dye or fiber may be colorfast to one factor but not another. Colorfast means that it holds onto color.
- **Crocking** is color transfer by rubbing. Imagine this scenario: dark blue denim against white shoes, or a white sofa. We have all experienced this unwanted color transfer. This is a much bigger problem for furniture or auto upholstery, as it is dye that is transferred, and most likely will not easily be removed.
- **Bleeding** results when a color runs when immersed in liquid. The dye that bleeds all over a load of laundry, for example, is called a fugitive dye and is simply excess dye that has not yet been rinsed out. This could be a problem when a fabric has a pattern with a bright color in it. If liquid is spilled on it, sometimes a bright or dark color will bleed onto the lighter areas.

Printed Fabrics

A print is color applied in a pattern or design. Prints have special considerations in placement and generally require more fabric in order to line up the elements of the design. The placement of the design elements must be evaluated for orientation and direction of the motifs.

The following are terms that the designer must become acquainted with, when ordering prints for residential or contract applications:

- **Ground** is the background color of the printed design. This may or may not be the dominant color of the design.
- **Colorway** is the specific combination of colors present in a design. The same design may be available in multiple colorways.
- **Repeat** is a single copy of a design. Sophisticated design repeats are often irregularly shaped to mask the repeat and to appear more organic.
- **Pattern match** relates to the layout of fabric pieces. It is imperative that the pieces of fabric line up so that the design or pattern flows across cushions and seams. More fabric is required for a printed fabric than a solid color. The larger the repeat size, the more yardage is required for a pattern match to occur. If a repeat is more than 25 inches, 40 percent more fabric will be required. This means that a 20 yard sofa will require an additional 8 yards—a significant added expense.
- **Railroading** is a process of cutting the fabric on the cross grain to minimize seams in large upholstered pieces. This process is usually more economical, as it maximizes the use of the fabric. Fabric is often referred to as railroaded when the pattern runs horizontal instead of vertical, encouraging railroading. Fabrics with a nap, such as velvet, should not be railroaded. A directional print, such as a stripe, will also have an impact on this decision.
- **Up the roll** is the opposite of railroading. This means that the fabric is used with the pattern running vertically, and the fabric is cut on the straight grain. Placing the fabric vertically is generally a better direction for strength. See Figure 15.2.

Prints can often be responsible for naming a fabric. Some fabrics get their name from their classic, recognizable printed patterns, such as those in Swatches 123 (ditsy or calico), 124 (paisley), or even 125 (toile de Jouy).

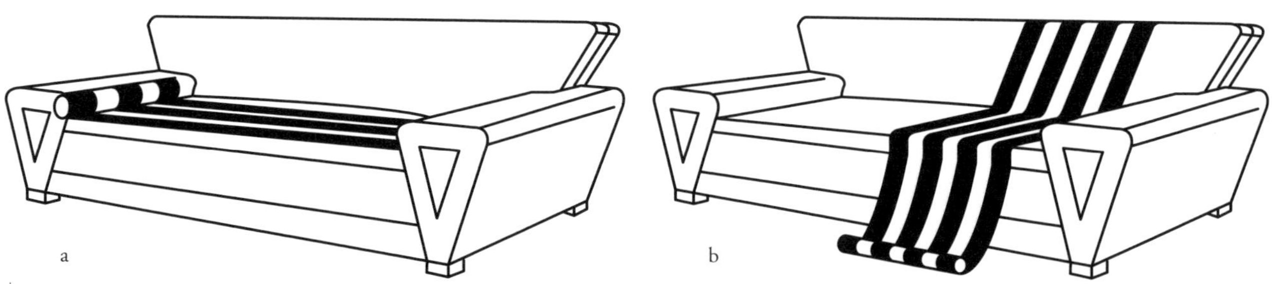

Figure 15.2 (a) Fabric used horizontally "railroaded" and (b) fabric used vertically "up the bolt"

If a printed pattern is not classically recognizable, then we identify it by default types of prints, such as direct print (printed without a background), blotch print (printing the background), or overprinting (printing on a previously dyed fabric). See Swatches 126–128.

Prints named for their print processes include a discharge print, flock print, and burnout. See Swatches 129–133.

A duplex print has been printed on both sides of the fabric, front and back. As in the swatch presented, they are often coordinating patterns. This one is a good choice for window treatments, as one side can face the outside world, and one is reserved for your own personal space (Swatch 134).

Finally, the resist prints straddle the dye and print processes. They sit in the print category because a pattern is created by their resist processes, but they are actually immersed in a dye bath.

- **Tie-dye** is a fabric resist, which means that the fabric is bound up to selectively resist the color and dropped into the dye bath. The pattern is created by the binding and can be quite beautiful.
- **Batik** is a wax resist. The wax coats areas of the fabric, prohibiting the dye from reaching those areas and creating lovely crackled designs where the wax cracks and the dye gets in.
- **Ikat** is a warp resist. The individual yarns of the fabric are tie-dyed in advance of fabric construction. Then the dyed yarns are reordered on the loom and woven together to create the design. The end result is that the images on the fabric have lovely feathered edges due to tensioning and shifting of the yarns.

In all cases, designs made from the resists are the same on the front and the back of the fabric, since they are immersed in dye baths. See Swatches 135–137.

[Reference Swatches 123–137]

Activity 15.1 Coloration Methods

Name: ____ Date: __________

Objective: To identify stage of dyeing or type of print
Dye class: Relates to fiber content.
Stage of dyeing: Choose from stock, yarn, piece, bleach, or print.
Prints: Identify the type of print. Choose from direct, blotch, overprint, burnout, or discharge.

Swatch	Fabric Name	Fiber Content	Probable Stage of Dyeing or Type of Print	Dye Class—Disperse, Reactive, or Acid
5	Jacquard	Cotton		
9	Burlap	Jute		
11	Melton	Wool		
16	Shantung	Dupioni silk		
20	Satin	Acetate		
67	Sateen	Cotton		
91	Terry	Cotton		
101	Fleece	Polyester		
108	Raschel lace	Nylon		
114	Oilcloth	PV/poly/cotton		
127	Linen look	Flax/rayon		

CHAPTER SIXTEEN

Fabrics Defined by Finishes

Reference Swatches

138. Flannel | 139. Flannelette | 140. Glazed Chintz
141. Moiré Taffeta | 142. Embossed | 143. Pleated
144. Suede Cloth | 145. Moleskin

Definition: A finish is anything done to a fiber, yarn, or fabric to change the way it looks, feels, or performs.

Finishes are achieved by either chemical or mechanical means, and just like dyes and prints, they are usually applied by converters. An example of a chemical finish is a flame-retardant chemical. Mechanical finishes include brushing the fabric to enhance its softness. Finishes are further divided into three basic groups: general, aesthetic, and functional.

General Finishes

All fabrics are subjected to general finishes; in fact, most fabrics have undergone at least a dozen general finishes by the time they reach the hands of the consumer. General finishes include such things using as washing, ironing, singeing, or bleaching the fabric.

Aesthetic Finishes

Aesthetic finishes change the way the fabric looks or feels, such as a moiré finish, pleating, or embossing. These are visible finishes and are often responsible for naming the fabric. Several are presented in this chapter.

Select Visible Aesthetic Finishes

Stiffening is applied to fabrics such as organza and organdy; if these fabrics were not stiffened, they would have other names (Swatches 40 and 41).

Napping is brushing the surface of the fabric with a wire brush, much like a pet hairbrush. The process changes the look, feel, and performance of the fabric. Flannel and flannelette (Swatches 138–139) have been napped, but so have flannel-back satin (Swatch 66) and fleece (Swatch 101). Velveteen, corduroy, velvet, and velour have all been sheared first, then napped. See Swatches 92–96 and 128. What is interesting is that all of these fabrics get their names and performance characteristics from their finishing process, yet each of them has a different name.

Calendering means to pass the fabric through heated rollers. When these rollers are dimensional, embossing occurs, such as in Swatch 142. This fabric has a pattern embossed onto the surface of the bonded sliver knit. Moiré is an embossed design with a specific woodgrain pattern, such as on the moiré taffeta that is Swatch 141.

Pleating (Swatch 143) is another type of calendering, accomplished by heat setting.

Sueding, actually called emerizing, is an important finish for many of the faux suedes prevalent in the marketplace today. Sandpaper is used on the surface of the fabric, which softens it without making it fuzzy. A faux suede is often more desirable than an actual suede due to its inherent durability (not to mention sparing the animal's life). Fabrics, such as sueded tricot (Swatch 105), suede cloth (Swatch 144), and moleskin (Swatch 145), are but a few examples of fabrics that have this finish today.

Functional Finishes

Functional finishes change the performance of the fabric and are often invisible. The consumer usually has to take these claims on faith, since they cannot be seen or felt. These include finishes such as stain resistance, antistatic, antimicrobial, flame retardancy (Swatch 113), or mothproofing, each of which is so important in the interior design world. Often these finishes determine whether the fabrics are more appropriate for residential or commercial purposes.

Emerizing/Sueding

There are few finishes where a change is made to the structure of the fiber and the effect is permanent. Many are durable (lasting 30–50 washes), which is, effectively, the life of the product. More commonly, finishes fall into the semi-durable category, which offers 10–20 washes before depletion. A temporary finish lasts through only one wash. Temporary finishes include a silicone finish, which washes out quickly, but initially gives the fabric a very soft hand. Finally, a renewable finish is one that you or the dry cleaner adds every time the item is cleaned. An example of a renewable finish would be fabric softener sheets, which soften and deodorize the fabric and add static resistance. Static resistance is an important, albeit temporary, finish. Most finishes, including a flame-retardant finish, will wash out over time.

[Reference Swatches 138–145]

Table 16.1 Fabric Selection by Application

Bedding/Bath/ Tableware	Window Treatment: Drapes	Window Treatment: Curtains	Upholstery	Accessories: Pillows, trims
Bamboo jersey	Bengaline	Antique satin	Bengaline	Antique satin
Bird's-eye piqué	Brocade	Batik	Brocade	Batik
Brocade	Brocatelle	Bird's-eye piqué	Brocatelle	Bengaline
Brushed tricot	Crepe	Burnout batiste	Canvas	Brocade
Cloqué	Crushed velvet	Burnout velvet	Cloqué	Brushed tricot
Cotton jersey	Damask	Canvas	Corduroy	Burlap
Damask	Ditzy print	Chambray	Crepe	Canvas
Dobby	Dobby	Clip spot	Crushed velvet	Clip spot
Double weave	Extra-yarn weave	Crepe de chine	Damask	Cloqué
Embossing	Flamestitch	Crushed filament voile	Denim	Crepe de chine
Fleece	Flock print	Crushed satin	Dobby	Crushed satin
Flannel-back satin	Flocked dotted swiss	Dimity	Double weave	Damask
Flannelette	Herringbone	Dobby	Drill	Dimity
Linen	Linen	Dotted swiss	Embossed velour	Dobby
Matelassé	Liseré	Duplex	Extra-yarn weave	Dotted swiss
Mattress pad	Paisley	Embroideries	Felt	Double weave
Micromodal jersey	Pincord	Filament voile	Flamestitch	Embossed velour
Minky	Pintucked taffeta	Gauze	Glazed chintz	Fleece
Moiré taffeta	Poplin	Gingham	Glen plaid	Ikat
Pocket cloth	Sateen	Ikat	Goose-eye twill	Knit velvet
Point d'esprit	Suede cloth	Linen	Grospoint	Linen
Poplin	Sueded tricot	Marquisette	Herringbone	Matelassé
Quilted	Taffeta	Moiré taffeta	Houndstooth	Minky
Raschel	Tapestry	Moleskin	Leather	Moiré taffeta
Raschel crochet	Toile de Jouy	Monk's cloth	Linen	Organdy
Sateen	Velvet	Organdy	Liseré	Pincord
Satin	Velveteen	Organza	Matelassé	Pintucked velvet
Seersucker	Voided velvet	Pleated	Moleskin	Point d'esprit
Suede cloth		Point d'esprit	Ottoman	Poplin
Terry cloth		Poplin	Pincord	Raschel
Tricot		Raschel	Pintucked taffeta	Raschel crochet
Tufted chenille		Raschel crochet	Pleather	Sateen
Tulle		Satin	Quilted	Satin
Waffle cloth		Seersucker	Repp	Seersucker
		Shantung	Sateen	Shantung
		Taffeta	Suede cloth	Sliver knit
		Tie-dye	Sueded tricot	Suede cloth
		Venise	Taffeta	Sueded tricot
		Voile	Tapestry	Taffeta
		Waffle cloth	Ticking	Tie-dye
			Toile de Jouy	Tricot
			Tweed	Tufted chenille
			Twill plaid	Tulle
			Ultrasuede	Velvet
			Velvet	Voided velvet
			Velveteen	
			Voided velvet	

Activity 16.1 Client Project

Name: ______________________________ Date: ____________

Project: You will be provided an assignment for a specific client by your instructor. (Example: Family room for Gen X couple with three children, Sweet 16 girl's bedroom, or a great room for a 40-ish couple that enjoys entertaining.) Design a room for this client and obtain fabrics for this room. You will evaluate these fabrics in terms of formal properties, followed by tabletop tests and relevant conclusions.

Layout:

- Draw a layout of the room you intend to design (scale is not required).
- Indicate the light aspect as it relates to the window to be covered with fabric (N, S, E, or W).
- Draw the placement of all the furniture in the room.
- Identify all elements (furniture and window treatments) in the room.
- Indicate which piece of furniture and which window you are designing.

Written content:

- State the number of persons in the household.
- State the ages and genders of all persons in the household.
- Describe the exact room to be designed.
- State the specific usage of the room by all family members.

Fabric:

- Purchase two coordinating fabrics for this room as follows:
    - 1 heavier fabric appropriate for upholstery
    - 1 sheer for a window treatment

Notes on Fabric Selection:

- Fabrics should coordinate in color.
- Each fabric should be of a **different fiber and fabric structure**.
- Each should be **100 percent** of a specific fiber. No blends.
- One fabric should be natural and one man-made.
- **Fabrics should be appropriate to their intended uses.**

Swatch Preparation:

- Cut each swatch 4" x 4", attached at the top only.
- Mount each swatch with the warp in the vertical direction.

Name: ______________________________ Date: ____________

- Identify the following for each swatch:
  - Fiber (burn to identify)
  - Yarn Structure (include evidence of any novelty yarn)
  - Fabric Name
  - Count
  - Coloration
  - Weight (show the math)
  - Obvious Finishes

Fabric Analysis

For each fabric provided, and using formal language, state three properties that justify this fiber/fabric as the **best choice** for your upholstery or drapery. Define these properties, and explain why this property is relevant to your client and their use of this fabric. No negatives.

Example: Cotton is hydrophilic, which means it is absorbent, and therefore easy to clean.

Notes on Presentation

- This should be a professional presentation, with typed content.
- Present each swatch on a separate page along with its complete identification (fiber, yarn, fabric, and count) and properties.

Table 9.2 Quick Reference Guide for the Three Basic Weaves

Plain Weaves 1:1	
Voile	Lightweight, slightly transparent, and made with hard-twisted (voile-twist) yarns, this plain-weave fabric has a slightly crisp, grainy hand. In filament form, this fabric is excellent for sheers. It is sometimes crushed for a decorative effect. Can also be made of cotton.
Organdy	Lightweight, transparent, plain-weave, stiffened cotton for maximum body.
Organza	Made of a trilobal synthetic for maximum luster, this plain-weave fabric has been stiffened for a lot of body. Better ones are made of silk.
Gauze	Gauze is a cotton-like, plain-weave fabric made with a low count, which is what causes the fabric to appear wrinkled.
Gingham	A cotton-like, lightweight, plain-weave fabric made with alternating blocks of white and colored yarns, creating a checked design.
Madras	A cotton-like, lightweight, plain-weave fabric made with several colored yarns, creating a plain-weave, plaid fabric.
Chambray	A cotton-like, lightweight, plain-weave fabric made with colored yarns in one direction and white yarns in the other.
Shantung	Made with a filament warp and slub weft in a plain-weave fabric; dupioni silk is usually made into a shantung.
Butcher linen	A plain-weave fabric made of bast fibers, usually flax, but also ramie or hemp. *Butcher linen* means that it is fairly heavyweight.
Basket Weaves 1:1, 2:2, and so on	
Monk's Cloth	A large, sometimes decorative, cotton, basket weave, usually found in a 4:4 ratio.
Canvas	A heavyweight basket weave, appropriate for upholstery. Used both indoors and out.
Rib Weaves 1:1	
Taffeta	A fine filament, this horizontal rib is a good choice for more formal window treatments and accessories due to its crispness and body.
Iridescent taffeta	*Iridescent* means that it is made with colored yarns that are opposite on the color wheel, causing the fabric to shift colors.
Poplin	This rib-weave fabric, made of cotton, is a good choice for drapery lining. Broadcloth is a slightly lighter weight counterpart to poplin.
Repp	Repp is a large, decorative, and sturdy rib.
Bengaline	Made with a filament warp and a heavier-spun weft, this is a rather flexible rib. Lustrous if made of acetate.
Ottoman	Ottoman is a larger, occasional, or more varied rib than bengaline.
Crepe de chine	A fine horizontal rib weave made of S & Z yarns with slight transparency and luster; this fabric is known for its drape.
Pincord	A fine cotton, vertical rib, this fabric is made with alternating white and a color warp with a white weft, creating a vertical stripe.
Dimity	Dimity is a lightweight, cotton fabric, appropriate for light curtains. It incorporates a vertical, occasional rib.

(*Table continued on next page*)

Table 9.2 Quick Reference Guide for the Three Basic Weaves *(continued)*	
Twills 2:1, 3:1, 2:2	
Denim	Denim is a 2:1 or 3:1 cotton twill made with an indigo-colored warp and a white weft.
Drill or bull denim	The heaviest of the twills, drill is a solid-color cotton fabric weighing 10–18 oz., with canvas-type applications.
Ticking	Ticking is a heavier cotton twill used to cover mattresses and pillows. Today it has been adapted to decorative upholstery purposes.
Herringbone	A reversing twill, this fabric can be yarn, piece, or stock dyed. This even-sided twill is a very strong dynamic design.
Houndstooth	Houndstooth is made with two or more colors and is a broken twill; the appearance is that of a jagged hound dog's tooth. Even-sided.
Glen plaid	Glen plaid is a yarn-dyed plaid made with multiple colors, with a houndstooth in it.
Twill plaid	Twill plaid is an even sided, 2:2 twill. The fabric is often napped and turned into a flannel.
Satins 4:1	
Satin	Satins, in general, have high luster and a lot of body.
Charmeuse	The lightest of the satins, the fabric is extremely drapey with a lot of luster.
Antique satin	A satin weave made with slub yarns in the weft.
Flannel-back satin	This satin weave is soft and fuzzy on the back side, lustrous on the face.
Sateen	Made of cotton, this satin has the lowest luster and is appropriate for many upholstery applications.

Table 10.1 Quick Reference Guide for Complex and Pile Weaves

Complex Weaves	
Seersucker	This cotton-like fabric has vertical, puckered stripes and can be yarn dyed, piece dyed, or printed.
Dobby	This cloth is made of two weaves creating a small, geometric, structural design.
Bird's-eye piqué	This is a cotton-like, two-dimensional, dobby design, with horizontal floats. The pattern resembles little eyes.
Waffle weave	This is a cotton-like, two-dimensional, dobby design, with floats in both directions. The pattern resembles a sheet of waffles.
Crepe	Also a dobby, this fabric has the appearance of a random structure, maximizing texture and drape.
Extra-yarn weave	Uses an extra set of yarns (warp or weft) to create a design. The yarns float along the back side of the fabric until needed again.
Clip spot	This is an extra-yarn weave, but the floats have been clipped off after weaving.
Dotted swiss	Dotted swiss is a type of clip spot; the pattern is made up of small dots, or spots.
Tapestry	Tapestry is made of a single weave (usually plain or rib weave) made of many colors, creating a pictorial image. Simplest structure, heaviest weight, with many colors and imagery.
Filament damask	Damask is a type of jacquard that uses satin weaves to create a pattern. The designs are mirror image front to back (which means the color or luster reverses on the back side). This one is made of filament yarns.
Tablecloth damask	Same as filament damask, but made of cotton or spun yarns.
Brocade	Brocade is the most complicated in structure and color. Made of several weaves and many colors. Not reversible. Liseré is a type of French brocade that has vertical stripes and flowers.
Brocatelle	Same as above but heavier and made of spun yarns.
Double weave	Two independent fabrics, woven together. Face and back can be remarkably different. Can have four or five sets of yarns. Pocket cloth is a double weave that has enclosed pockets in the structure.
Double-sided satin	Two-faced satin, which means that the lustrous surface is on both sides.
Matelassé	A double weave where one layer shrinks causing the other to pucker, creating a dimensional surface.
Cloqué	A double-weave structure that has some padding or extra yarns in the center to create a dimensional surface. The fabric appears quilted.
Leno weave	A lacey, casement-weight fabric created when the warp yarns create a figure 8 around the weft yarns. This creates a low-count structure that is exceptionally stable.
Pile Weaves	
Terry cloth	Cotton-like fabric with loops on the surface of both sides, often made into towels.
Terry velour	Same as above but loops are sheared on one side.
Velveteen	Cotton-like fabric; the loops have been sheared. Low luster. This is a velvet made of cotton.
Corduroy	Velveteen made in rows. Rows could be thin, as in pinwale corduroy, or thick, as in widewale corduroy.
Velvet	A pile weave made of lustrous filament yarns that is sheared and napped for maximum luster and drape.
Panné velvet	The velvet has been pressed down in a single direction to increase luster and drape even further. Smooth in one direction, sharp in the other.
Crushed velvet	The velvet has been crushed for maximum texture.
Grospoint	A heavy and sturdy upholstery fabric with rows of large uncut loops.

Table 12.3 Quick Reference Guide for Weft Knits

Weft Knits	
Jersey knit	Front and back of this fabric is different. Fabric curls to the face. Wales on the front, courses on the back. 3–18 oz.
Fair isle/jersey jacquard	Uses floats and colored yarns to make a complex pattern or image.
Rib Knits	
Rib knit 1 × 1	Made of alternating knit and purl stitches. Rib knits do not curl.
Rib knit 2 × 2	Two Vs in each rib.
Pointelle	Rib knit with tuck stitches creating a small pattern like hearts or stars.
Thermal knit	Rib knit with tuck stitches, also known as *waffle knit*.
Cable knit	Rib variation with plaits or braids on one side.
Double-Weft Knits	
Interlock	Looks like 1 × 1 rib; generally lighter weight than rib knits.
Double-knit jacquard	Complex—two layers, yarn dyed. Front and back are different.
Ponte di roma	Little stretch, looks like a double piqué. Very stable.
Double-knit matelassé	Double knit with a blistered, dimensional surface.
Pile Knits	
Knitted terry	Knit terry cloth.
French terry	Lower, flatter loops. Better markets.
Knit velour	Cut loops—equivalent to a knit velveteen.
Sliver/high pile knit	Fake fur. Can be long hair or short.
Crushed knit velvet	The pile has been crushed, creating lots of texture.
Panné knit velvet	Pile laid in one direction. Highest luster, best drape.
Sweatshirt fleece	Napped on one side. Single-sided fleece.
Fleece	Napped on two sides. Two-sided fleece.

Table 13.3 Quick Reference Guide for Warp Knits

Tricots	
Tricot	Wales on the front, courses on the back—little stretch. Stable.
Brushed tricot	Napped tricot. Soft on one side.
Sueded tricot	Emerized or sanded tricot. Suede-like. A common type of imitation suede.
Satin tricot	Smooth and lustrous on one side. Little stretch in width.
Sliver knit	Fake fur on a tricot body. Stable enough for upholstery.
Minky	A very soft, short hair, alternative to fake furs. Used in children's blankets.
Raschels	
Hex net	Hexagonal grid. Soft and flexible.
Tulle	Netting or mesh. Small hex or diamond-shaped grid that has been stiffened.
Raschel laces	Knit lace on tulle or hex-net ground. Most laces are raschels.
Cut press	Raschel lace with vertical, zigzag patterning.
Crochet raschel	Machine-made crochet, often cotton-like.
Point d'esprit	Looks like a knit version of a dotted swiss.

Swatch 1

Fiber content: Raw cotton
(S) (R) (B)

Swatch 2

Fiber content: Cotton
Yarn construction: Two-ply spun
Yarn size: 8/2 Yarn count
(S) (R) (B)

Raw Fiber

Swatch 1

Yarn

Swatch 2

(B) = biodegradable, (C) = Closed Loop, (R) = Recyclable, (S) = Sustainable

Yarn

Swatch 3

Fabric

Swatch 4

Swatch 3

Fiber content: Polyester
Yarn construction: Multifilament
Yarn size: 35 tex or 310 denier
Ⓡ

Swatch 4

Fiber content: Cotton
Yarn construction: Spun
Fabric name: Muslin
Ⓢ Ⓡ Ⓑ

Swatch 5

Fiber content: Cotton
Yarn construction: Spun
Fabric name: Jacquard
Count: 100 × 72
Coloration: ________________
Weight: ________________
Uses: Curtains, tableware
Ⓢ Ⓡ Ⓑ

Swatch 6

Fiber content: Organically colorgrown cotton
Yarn construction: ________________
Fabric name: ________________
Count: 104 × 48
Coloration: Colorgrown; natural, undyed
Weight: ________________
Uses: Window treatments, tableware
Made in USA
Ⓢ Ⓡ Ⓑ

Cellulose Fibers

Swatch 5

Cellulose Fibers

Swatch 6

Cellulose Fibers

Swatch 7

Cellulose Fibers

Swatch 8

Swatch 7

Fiber content: Flax
Yarn construction: Spun
Fabric name: "Vintage" linen
Double weave
Count: 96 × 92
Coloration: Piece dyed
Weight: 11.32 oz./yd.^{2}
Uses: Windows, accessories, tableware
Finishes: Distressing

Swatch 8

Fiber content: Hemp
Yarn construction: Warp—Two-ply spun
Weft—Single spun
Fabric name: Hemp linen
Count: 24 × 18
Coloration: Piece dyed
Weight: 13.87 oz./yd.^{2}
Uses: Windows, accessories, tableware

Swatch 9

Fiber content: Jute
Yarn construction: Spun
Fabric name: Burlap
Count: 11 × 11
Coloration: ______________
Weight: 9.76 oz./yd.^{2}
Uses: Accessories, rugs, baskets

Swatch 10

Fiber content: Italian raffia
Yarn construction: Warp—Single spun
Weft—Two-ply spun
Fabric name: Bonded herringbone
Tricot backing
Count: 12 × 8
Coloration: Piece dyed
Weight: 23.27 oz./yd.^{2}
Uses: Outdoor furniture, rugs, baskets

Cellulose Fibers

Swatch 9

Cellulose Fibers

Swatch 10

Protein Fibers

Swatch 11

Swatch 12

Swatch 11

Fiber content: Wool
Yarn construction: Spun
Fabric name: Melton
Count: 34 × 24
Coloration: ________________
Weight: 12.04 oz./yd.^{2}
Uses: Upholstery
Finishes: Fulling
Ⓢ Ⓡ Ⓑ

Characteristics

Melton is a densely woven, partially felted fabric with excellent abrasion resistance due to the wool.

Swatch 12

Fiber content: Mohair
Yarn construction: Spun
Fabric name: Velvet
Count: N/A
Coloration: Piece dyed
Weight: 29.97 oz./yd.^{2}
Uses: Upholstery
Finishes: Shearing, napping
Ⓢ Ⓡ Ⓑ

Characteristics

Mohair velvet is an old, luxury, upholstery selection. It has long been a quality standard for theater seating.

Swatch 13

Fiber content: Recycled silk
Yarn construction: ________________
Fabric name: Novelty rib
Count: 66 × 38
Coloration: Yarn dyed
Weight: 8.73 oz./yd.^{2}
Uses: Light upholstery, pillows, accessories

Characteristics

This silk fabric is made from deconstructed, stripped, and re-woven Indian sari cloth. As such, the fabric has varied texture and a lot of energy.

Swatch 14

Fiber content: Wild silk
Yarn construction: ________________
Fabric name: ________________
Count: 56 × 56
Coloration: Yarn dyed
Weight: 4.92 oz./yd.^{2}
Uses: Light upholstery, pillows, accessories

Characteristics

Wild silk is made from broken cocoons, or other silk caterpillars (not Bombyx Mori, which makes the fine, shiny silk fabrics). The surface of this fabric is duller and not as silky, as the yarns are spun instead of filament.

Protein Fibers

Swatch 13

Protein Fibers

Swatch 14

Swatch 15

Swatch 15

Fiber content: Silk
Fabric name: Silk noil
Yarn construction: Spun
Count: 53 × 41
Coloration: Greige goods
Weight: 4.04 oz./yd.^{2}
Uses: Pillows, Accessories

Characteristics

Silk noil is waste silk, with a very cotton-like hand due to its spun yarns. Notice that the surface is slightly sticky.

Protein Fibers

Swatch 16

Swatch 16

Fiber content: Silk dupioni
Fabric name: Shantung
Yarn construction: Warp—Filament
Weft—Slub
Count: 108 × 72
Coloration: ______________
Weight: 2.74 oz./yd.^{2}
Uses: Decorative, pillows, accessories

Characteristics

Dupioni silk comes from two caterpillars that are joined together in the spinning process. Because of the joining, the fabric has both slubs and the rich luster of cultivated silk.

Swatch 17

Fiber content: Face—Rayon
Backing—Polyester

Yarn construction: Filament

Fabric name: Bonded—Panné velvet (face)
Jersey (back)

Count: 112 × 62 on face

Coloration: Piece dyed

Weight: 10.85 oz./yd.^{2}

Uses: Light upholstery, windows, pillows

Finishes: Shearing, napping

Characteristics

Rayon was originally made from the pulp of a mulberry tree in an attempt to create "artificial silk." The fiber is often chosen for its excellent drape and soft hand.

Swatch 18

Fiber content: Bamboo rayon

Yarn construction: Spun

Fabric name: 1 × 1 Rib knit

Count: 58 × 44

Coloration: Greige goods

Weight: 5.52 oz./yd.^{2}

Uses: Sheets

Ⓢ Ⓡ Ⓑ

Characteristics

Bamboo is currently considered a type of rayon (as it is made in the same process, but from a different tree—bamboo). It is fast growing, does not require pesticides, and is biodegradable. Bamboo is often processed with harsh chemicals, however. The fiber is known for its antibacterial qualities, along with its excellent wicking and a very soft hand.

Manufactured Cellulosic Fiber

Swatch 17

Manufactured Cellulosic Fiber

Swatch 18

Manufactured Cellulose Fiber

Swatch 19

Swatch 19

Fiber content: Micromodal®
Yarn construction: Spun
Fabric name: ______________
Count: ______________
Coloration: Piece bleached
Weight: 4.04 oz./yd.^{2}
Uses: Sheets
Ⓒ Ⓡ Ⓢ Ⓑ

Characteristics

Modal® is relatively new to sheeting. Made from beech trees, it is very lightweight, wicks well, is naturally antibacterial, and is one of the softest fibers. In microfiber form, it is called Micromodal®. It also is more dimensionally stable than rayon.

Manufactured Cellulose Fiber

Swatch 20

Swatch 20

Fiber content: Acetate
Yarn construction: Filament
Fabric name: ______________
Count: 158 × 104
Coloration: ______________
Weight: 3.66 oz./yd.^{2}
Uses: Pillows, bedding, window treatments
Ⓢ Ⓡ Ⓑ

Characteristics

Acetate is made of wood pulp and chemicals, so it burns and melts. The fiber is used for bedding (comforters and quilts) and decorative pillows due to its inherent luster. This example is iridescent due to the fact that it is made of two complementary colors.

Swatch 21

Fiber content: Ingeo® corn
Yarn construction: Filament
Fabric name: Plain weave
Count: 64 × 52
Coloration: Piece bleached
Weight: 4.88 oz./yd.^{2}
Uses: Drapery linings

Characteristics

Ingeo is a polymer derived from corn fibers. PLA is polylactic acid polymer and can be used to replace synthetics. It is biodegradable, hydrophilic, and has good shape retention and resiliency.

Manufactured Cellulosic Fiber

Swatch 21

Manufactured Mineral

Swatch 22

Manufactured Mineral

Swatch 23

Swatch 22

Fiber content: Glass
Yarn construction: Filament
Fabric name: ______________
Count: 64 × 60
Coloration: Natural
Weight: 3.05 oz./yd.^{2}
Uses: Industrial window treatments

Characteristics

Glass is the heaviest fiber and is mostly used for industrial functions, such as filters, drapes, and fiber optics. Because it is nonflammable, it can be an excellent choice.

Swatch 23

Fiber content: 72% Flax/19% Cotton/9% Metallic
Yarn construction: ______________
Fabric name: Metallic linen—Basket-weave variation
Count: 70 × 60
Coloration: Yarn dyed
Weight: 4.31
Uses: Decorative items, window treatments

Characteristics

Metallic fibers (yarns) are used for purely decorative purposes. Notice that metallics are used in a blend, since they are not strong enough to stand alone in a fabric. Metallics add sparkle to a design.

Metals (such as steel) are often added to reduce static in a carpet.

Swatch 24

Fiber content: Nylon
Yarn construction: Filament
Fabric name: Basket-weave webbing
Count: 38 × 36
Coloration: Print
Weight: 8.94 oz./$yd.^2$
Uses: Upholstery, tabletop

Characteristics

Nylon is extremely lightweight and strong. For this reason, it is a good choice for upholstery.

Synthetic Fibers

Swatch 24

Swatch 25

Fiber content: Acrylic
Yarn construction: ____________
Fabric name: Flannel
Count: 42 × 30
Coloration: Yarn dyed
Weight: 5.62 oz./$yd.^2$
Uses: Light upholstery
Finishes: Napping

Characteristics

Acrylic is the best wool substitute. It is lightweight, does not shrink, and is hypoallergenic and hydrophobic.

Synthetic Fibers

Swatch 25

Synthetic Fibers

Swatch 26

Swatch 26

Fiber content: Polyester
Yarn construction: Filament
Fabric name: Pintucked taffeta
Count: 144 × 80
Coloration: Yarn dyed
Weight: 3.54 oz./yd.²
Uses: Bedding, windows

Characteristics

Polyester is a chameleon in that it has many different appearances. It can imitate silk, cotton, and linen and all for less money. It is also the most common blended fiber. Polyester is a workhorse, with excellent dimensional stability, and is known for its resilience. It is an excellent choice for upholstery and bedding, as it repels stains.

Synthetic Fibers

Swatch 27

Swatch 27

Fiber content: Olefin
Yarn construction: Filament
Fabric name: Imitation linen ("Linen look")
Count: 42 × 25
Coloration: Solution dyed
Weight: 5.52 oz./yd.²
Uses: Outdoor upholstery, padded headboard

Characteristics

Olefin is often used outdoors, as it is extremely hydrophobic and incredibly durable.

Swatch 28

Fiber content: Aramid
Yarn construction: Spun
Fabric name: Ripstop
Count: 88 × 60
Coloration: Piece dyed
Weight: 6.71 oz./yd.^{2}
Uses: Upholstery

Characteristics

Aramid is the strongest fiber and has many practical uses, including fire protective gear. A good choice in high traffic areas where strength is essential.

Synthetic Fibers

Swatch 28

Swatch 29

Fiber content: Polyester
Yarn construction: Multifilament
Fabric structure: Jacquard
Fabric name: Brocade
Count: 200 × 116
Coloration: Yarn dyed
Weight: 5.37
Uses: Windows, pillows, accessories
Ⓡ

Characteristics

This fabric is a jacquard weave, specifically a brocade, made of filament yarns. Note the shiny surface of the fabric caused by the long filament yarns.

Swatch 30

Fiber content: Cotton
Yarn construction: Spun
Fabric structure: ______________
Fabric name: ______________
Count: 128 × 46
Coloration: Piece bleached
Weight: 7.84 oz./yd.^{2}
Uses: Pillows, accessories, light upholstery
Ⓢ Ⓡ Ⓑ

Characteristics

The fabric is a dobby, made of spun yarns. Note the soft, fuzzy surface of the fabric, created by spun yarns.

Filament Yarns

Swatch 29

Spun Yarns

Swatch 30

Two-ply Spun Yarns

Swatch 31

Novelty Yarn

Swatch 32

Swatch 31

Fiber content: Cotton
Yarn construction: Two-ply spun
Fabric structure: Jacquard
Fabric name: Tapestry
Count: 144 × 40
Coloration: Yarn dyed
Weight: 18.88
Uses: Upholstery
S R B

Characteristics

The yarns that make this tapestry are two-ply spun. Notice that although the yarns are different sizes, they each will untwist into two separate yarns, called singles.

Swatch 32

Fiber content: Acrylic
Yarn construction: Bouclé & spun
Fabric structure: Twill
Fabric name: Houndstooth bouclé
Count: 25 × 25
Coloration: Yarn dyed
Weight: 9.74 oz./yd.^{2}
Uses: Light upholstery, decorative blankets, pillows, accessories
R

Characteristics

Bouclé is a loop yarn that creates a lot of interest on the surface of the fabric. As the yarn has so much personality, the fabric is often named for this yarn type.

Swatch 33

Fiber content: Polyester
Yarn construction: Warp—filament
Weft—Chenille
Fabric structure: Plain weave
Fabric name: Plain-weave chenille
Count: 56 × 36
Coloration: Yarn dyed
Weight: 11.29 oz./yd.^{2}
Uses: Pillows, accessories, upholstery

Characteristics

Chenille is French for caterpillar, and the yarns resemble caterpillars. This novelty yarn type has a velvety texture and can be made of many fibers. Once again, the yarns are so dominant that the fabric is named for the yarn structure.

Swatch 34

Fiber content: Cotton/polyester
Yarn construction: Warp—Filament
Weft—Chenille
Fabric structure: Double-weave jacquard
Fabric name: Double-weave chenille tapestry
Count: 160 × 50
Coloration: Yarn dyed
Weight: 13.17 oz./yd.^{2}
Uses: Pillows, accessories, upholstery

Characteristics

Using chenille in a damask or a double weave is quite popular today. In this case the chenille yarns are made of cotton.

Novelty Yarn

Swatch 33

Novelty Yarn

Swatch 34

Swatch 35

Swatch 35

Fiber content: Wool
Yarn construction: Spun
Fabric structure: Plain weave
Fabric name: Tweed
Count: 30 × 12
Coloration: Stock dyed
Weight: 5.94
Uses: Pillows, accessories, light upholstery
S R B

Characteristics

Tweed can be made of many structures; the name comes from the stock dyed and slub yarns. The heathery, tweedy appearance caused by the stock-dyed yarns identifies it as a tweed. This one is a plain-weave tweed as opposed to a twill tweed or a herringbone tweed.

Swatch 36

Fiber content: Cotton
Yarn construction: Spun
Fabric structure: Plain weave
Fabric name: Voile
Count: 74 × 72
Coloration: Piece bleached
Weight: 1.73 oz./yd.^{2}
Uses: Window sheers
Ⓢ Ⓡ Ⓑ

Characteristics

Voile is French for veil, and there are three different voiles presented here for comparison. Notice the difference between the voile made of filament yarns and the voile made of spun yarns. This voile is made of hard-twisted spun cotton yarns, which maximizes transparency.

Swatch 37

Fiber content: Polyester
Yarn construction: Filament
Fabric structure: Plain weave
Fabric name: Filament Voile
Count: 74 × 68
Coloration: Piece bleached
Weight: 1.46 oz./yd.^{2}
Uses: Window sheers
Ⓡ

Characteristics

This is a filament variation of voile, which is a little stiffer and a little sheerer.

Plain Weaves

Swatch 36

Plain Weaves

Swatch 37

Swatch 38

Swatch 38

Fiber content: Polyester
Yarn construction: Filament
Fabric structure: Plain weave
Fabric name: Crushed voile
Count: 82 × 78
Coloration: Piece bleached
Weight: 1.9 oz./yd.^{2}
Uses: Window sheers
Finishes: Crushing
Ⓡ

Characteristics

This is a filament voile with a crushed finish to add texture and a little more privacy in the sheer.

Swatch 39

Swatch 39

Fiber content: Cotton
Yarn construction: ________________
Fabric structure: Plain weave
Fabric name: Gauze
Count: 28 × 24
Coloration: Piece dyed
Weight: 1.1 oz./yd.^{2}
Uses: Window sheers
Ⓢ Ⓡ Ⓑ

Characteristics

Gauze is another sheer made of a plain-weave cotton with an extremely low thread count. The thread count causes the fabric to shrink and wrinkle, giving it that characteristic disorderly appearance. A close relative of scrim, both fabrics are lightweight and plain weave—used for windows.

Swatch 40

Fiber content: Pima cotton
Yarn construction: Spun
Fabric structure: Plain weave
Fabric name: Organdy
Count: 100 × 72
Coloration: Piece bleached
Weight: 1.2 oz./yd.^{2}
Uses: Sheers

Characteristics

Organdy is a lightweight, plain-weave, cotton fabric, made with a stiffening finish. The finish makes it the sheerest of the cotton fabrics. It also has body not present in the previous sheers.

Swatch 41

Fiber content: Nylon
Yarn construction: ________________
Fabric structure: Plain weave
Fabric name: Organza
Count: 96 × 96
Coloration: Piece dyed
Weight: 0.73 oz./yd.^{2}
Uses: Sheers

Characteristics

Organza is the filament counterpart to organdy. It is both more sheer and more lustrous. This example is a glass organza, for maximum luster.

Plain Weaves

Swatch 40

Plain Weaves

Swatch 41

Plain Weaves

Swatch 42

Swatch 42

Fiber content: Cotton
Yarn construction: Spun
Fabric structure: Plain weave
Fabric name: Gingham
Count: 48 × 31
Coloration: Yarn dyed
Weight: 7.94 oz./yd.^{2}
Uses: Light upholstery, pillows, accessories
Ⓢ Ⓡ Ⓑ

Characteristics

Gingham is a cotton-like fabric with an even checkered pattern. The fabric is made of white (or off-white) and a color. Some gingham variations are smaller than this and some are substantially larger, called Buffalo check.

Plain Weaves

Swatch 43

Swatch 43

Fiber content: Cotton
Yarn construction: Spun
Fabric structure: Plain weave
Fabric name: Chambray
Count: 64 × 50
Coloration: Yarn dyed
Weight: 5.07 oz./yd.^{2}
Uses: Pillows, accessories, curtains
Ⓢ Ⓡ Ⓑ

Characteristics

Chambray is another linen or cotton-like fabric that uses white and a color in its yarn organization. Chambray has all white yarns in one direction and colored yarns in the other, creating a denim-like appearance.

Swatch 44

Fiber content: Polyester
Yarn construction: Warp—Filament
Weft—Filament & slub
Fabric structure: Plain weave
Fabric name: Iridescent shantung
Count: 124 × 70
Coloration: Yarn dyed
Weight: 3.2 oz./yd.^{2}
Uses: Window dressing, accessories

Characteristics

Shantung has filament yarns in the warp direction and slub yarns in the weft, creating a subtle surface texture. Iridescent is a prefix that can be added to many fabrics and suggests that the fabric is made of two complementary colors, causing the fabric to shimmer and change colors with movement. Iridescent shantung has both qualities.

Swatch 45

Fiber content: Flax
Yarn construction: Spun
Fabric structure: Plain weave
Fabric name: Crash or butcher linen
Count: 24 × 20
Coloration: Bleached
Weight: 11.06 oz./yd.^{2}
Uses: Upholstery, bedding, accessories

Characteristics

Butcher linen or crash is the heaviest of the linens, getting its name from butcher shop aprons, designed to be quite sturdy. Linen has slubs in both directions and must be made of a bast fiber to be called linen.

Plain Weaves

Swatch 44

Plain Weaves

Swatch 45

Swatch 46

Fiber content: Polyester
Yarn construction: ____________
Fabric structure: Basket weave
Fabric name: Monk's cloth
Count: 36 × 40
Coloration: Piece bleached
Weight: 5.08 oz./yd.^{2}
Uses: Curtains, upholstery backings

Characteristics

Monk's cloth is traditionally made of cotton. It is a lightweight, open basket weave, originally used to make monk's robes. This fabric is a good choice for lightweight curtains, accessories, or pillows.

Swatch 47

Fiber content: Polyester
Yarn construction: Filament
Fabric structure: Basket weave
Fabric name: Sunbrella® canvas
Count: 150 × 115
Coloration: Yarn dyed
Weight: 8.66 oz./yd.^{2}
Uses: Outdoor upholstery
Finishes: Water repellent

Characteristics

Canvas is a durable, all-purpose fabric, made in a basket weave, which promotes resilience and, in this case, strength. Particularly sturdy canvases are called duck.

Sunbrella® canvas incorporates a fluorocarbon finish that makes the fabric water repellent. It is also solution dyed for long-lasting colorfastness.

Basket Weaves

Swatch 46

Basket Weaves

Swatch 47

Rib Weaves

Swatch 48

Swatch 48

Fiber content: Acetate
Yarn construction: Filament
Fabric structure: Rib weave
Fabric name: Taffeta
Count: 176 × 52
Coloration: Piece dyed
Weight: 2.58 oz./yd.^{2}
Uses: Window treatments, accessories, bedding
Ⓢ Ⓡ Ⓑ

Characteristics

Taffeta can easily be identified by the rustling sound it makes. Run your fingernails down the surface of the fabric for confirmation. Taffeta is a good choice when body (not drape) is required.

Rib Weaves

Swatch 49

Swatch 49

Fiber content: Rayon/polyester
Yarn construction: Filament
Fabric structure: Rib weave
Fabric name: Crepe de chine
Count: 212 × 124
Coloration: Piece dyed
Weight: 3.73 oz./yd.^{2}
Uses: Window treatments, accessories

Characteristics

Crepe de chine is a good choice when drape is required. This fabric is made of alternating S and Z yarns in the weft direction. The fabric has a very fine hand and is slightly lustrous.

Swatch 50

Fiber content: Polyester/cotton
Yarn construction: Warp—Spun
Weft—Filament
Fabric structure: Rib weave
Fabric name: Broadcloth (Drapery lining)
Count: 82 × 44
Coloration: Greige goods
Weight: 3.0 oz./yd.^{2}
Uses: Drapery lining

Characteristics

Often fabrics need a lining to be sturdy enough or have enough body to hang properly. Broadcloth is a good choice for a drapery lining. The fabric could also be used to back a lighter weight upholstery fabric. This extra layer, sometimes called an interlining, can be added to some lighter weight upholstery fabrics like chenille. This layer assists in keeping the fabric from sagging or stretching out. The layer stabilizes lighter weight upholstery fabrics and improves wear. Broadcloth is traditionally a cotton fabric.

Swatch 51

Fiber content: Cotton
Yarn construction: Spun
Fabric structure: Rib weave
Fabric name: Repp
Count: 64 × 36
Coloration: Yarn dyed
Weight: 10.1 oz./yd.^{2}
Uses: Upholstery, drapes

Characteristics

Repp is a durable, hard-working upholstery fabric with a lovely, horizontal ribbed surface texture.

Rib Weaves

Swatch 50

Rib Weaves

Swatch 51

Rib Weaves

Swatch 52

Swatch 52

Fiber content: Polyester

Yarn construction: Warp—Filament
Weft—Spun

Fabric structure: Rib weave

Fabric name: Bengaline

Count: 132 × 32

Coloration: Piece dyed

Weight: 4.92 oz./yd.²

Uses: Windows, pillows, accessories

Characteristics

The name bengaline comes from the Bengal region of India. The fabric has a thicker weft and thinner warp, creating a strong horizontal rib. Note the unbalanced count that assists in creating the rib structure.

Rib Weaves

Swatch 53

Swatch 53

Fiber content: Cotton

Yarn construction: Spun

Fabric structure: Rib weave

Fabric name: Ottoman

Count: 104 × 68

Coloration: Piece dyed

Weight: 8.87 oz./yd.²

Uses: Upholstery, pillows, accessories

Characteristics

Ottoman is a larger horizontal rib. The fabric could have various sized ribs or constructed as an occasional rib, like the one presented here. An occasional rib has spaces between the ribs.

Swatch 54

Fiber content: Cotton
Yarn construction: Spun
Fabric structure: Rib weave
Fabric name: Pincord
Count: 68 × 44
Coloration: Yarn dyed
Weight: 4.27 oz./yd.^{2}
Uses: Light upholstery, pillows, accessories

Characteristics

Pincord is a vertical rib. Note that this is not just yarn dyed, but there is an actual vertical rib that creates texture and durability in the fabric. Pincord is always yarn dyed with alternating white and colored yarns, creating the vertical stripe aesthetic.

Vertical Rib Weave

Swatch 54

Swatch 55

Fiber content: Cotton
Yarn construction: Spun
Fabric structure: Rib weave
Fabric name: Dimity
Count: 94 × 68
Coloration: Piece bleached
Weight: 2.12 oz./yd.^{2}
Uses: Curtains

Characteristics

Dimity is a lighter, vertical rib, usually made of cotton. This occasional rib provides a great window curtain aesthetic.

Vertical Rib Weave

Swatch 55

Swatch 56

Fiber content: Cotton/spandex
Yarn construction: Spun
Fabric structure: Twill
Fabric name: Denim
Count: 80 × 42
Coloration: Yarn dyed
Weight: 10.4 oz./yd.^{2}
Uses: Upholstery, accessories

Characteristics

Denim was originally developed in the city of Nimes, France (de Nimes, meaning of Nimes). Traditionally, the fabric is made with a white warp and indigo dyed weft in a twill weave, but polyester, spandex, and other fibers are finding their way into denim fabrics.

Swatch 57

Fiber content: Cotton
Yarn construction: Spun
Fabric structure: Twill
Fabric name: Drill
Count: 62 × 42
Coloration: Piece dyed
Weight: 12.03 oz./yd.^{2}
Uses: Upholstery

Characteristics

Drill is a heavy and strong twill good for upholstery, bedding, and other accessories. Drill is a heavyweight, (above 10 oz.) solid-colored twill made of cotton, with a denim hand.

Twill Weave

Swatch 56

Twill Weave

Swatch 57

Swatch 58

Swatch 58

Fiber content: Cotton
Yarn construction: Spun
Fabric structure: Twill
Fabric name: Ticking
Count: 62 × 40
Coloration: Yarn dyed
Weight: 7.38 oz./yd.^{2}
Uses: Mattress & pillow covers, upholstery
S R B

Characteristics

Ticking was traditionally used as mattress or pillow covers, but it has been extended into upholstery for that "shabby chic" look. This cotton twill is quite durable.

Twill Weaves

Swatch 59

Swatch 59

Fiber content: Wool/polyester
Yarn construction: Spun
Fabric structure: Twill
Fabric name: Herringbone
Count: 42 × 32
Coloration: Yarn dyed
Weight: 8.61 oz./yd.^{2}
Uses: Upholstery

Characteristics

Herringbone is a reversing twill that gets its name from the resemblance to the bones of a herring fish. This is a strong twill suitable for accessories, pillows, or upholstery, and it can be found yarn dyed, piece dyed, or printed.

Swatch 60

Fiber content: Wool
Yarn construction: Spun
Fabric structure: Twill
Fabric name: Houndstooth
Count: 32 × 28
Coloration: Yarn dyed
Weight: 7.0 oz./yd.^{2}
Uses: Upholstery, blankets

Characteristics

Houndstooth is a broken twill that resembles a slanted four pointed star. The jagged edges are suggestive of a hound dog's teeth.

Swatch 61

Fiber content: Wool/polyester
Yarn construction: Spun
Fabric structure: Twill
Fabric name: Glen plaid
Count: 40 × 40
Coloration: Yarn dyed
Weight: 8.45 oz./yd.^{2}
Uses: Upholstery, blankets
Finishes: Napped

Characteristics

Glen plaid is a type of yarn dyed plaid that has houndstooth in it. As such, it has all the benefits of houndstooth.

Twill Weaves

Swatch 60

Twill Weaves

Swatch 61

Swatch 62

Swatch 62

Fiber content: Cotton
Yarn construction: Two-ply spun
Fabric structure: Twill
Fabric name: Chevron plaid
Count: 28 × 20
Coloration: Yarn dyed
Weight: 12.57 oz./yd.^{2}
Uses: Upholstery

Characteristics

There are many variations of twill plaids; the one presented here is a chevron. (The points in a chevron meet, while the points in a herringbone are staggered). Plaid's popularity is reinforced by the strength of a twill fabric. Other varieties include the two flannels in this text, Swatches 25 & 139.

Swatch 63

Fiber content: Polyester
Yarn construction: ________________
Fabric structure: Satin
Fabric name: Satin
Count: 264 × 80
Coloration: Piece dyed
Weight: 4.0 oz./yd.^{2}
Uses: Accessories, pillows, draperies

Characteristics

Satin is the most lustrous fabric of the three basic weaves, along with having the best drape, which is why it can be a good choice for drapes. Originally made of silk, satins today are made of polyester, nylon, rayon, and any other filament yarn.

Swatch 64

Fiber content: Polyester
Yarn construction: Filament
Fabric structure: Satin
Fabric name: Iridescent crushed satin
Count: 230 × 82
Coloration: Yarn dyed
Weight: 3.01 oz./yd.^{2}
Uses: Accessories, pillows, draperies
Finishes: Crushing

Characteristics

This satin is very similar to Swatch 63 except that it has a crushed finish, adding texture to the surface. Once again, the fabric is iridescent due to the complementary colored yarns.

Satin Weaves

Swatch 63

Satin Weaves

Swatch 64

Satin Weaves

Swatch 65

Swatch 65

Fiber content: Rayon

Yarn construction: Warp—Filament
Weft—Two-ply spun & slub

Fabric structure: Satin

Fabric name: Antique satin

Count: 196 × 50

Coloration: Piece dyed

Weight: 5.33 oz./yd.²

Uses: Windows, accessories, pillows, bedding

Ⓢ Ⓡ Ⓑ

Characteristics

Antique satin (sometimes called satin-back shantung) is a little heavier and more textured due to the slub yarns in the horizontal direction.

Satin Weaves

Swatch 66

Swatch 66

Fiber content: Polyester/cotton

Yarn construction: Warp—Filament
Weft—Spun, low twist

Fabric structure: Satin

Fabric name: Flannel-back satin

Count: 232 × 60

Coloration: Piece dyed

Weight: 5.92 oz./yd.²

Uses: Windows, accessories, pillows, bedding

Finishes: Napping

Characteristics

Flannel-back satin makes an excellent sheeting fabric; smooth and lustrous on the outside, and warm and cozy on the inside. Cotton has been added so that the fabric can be napped on the back side.

Swatch 67

Fiber content: Cotton
Yarn construction: Spun
Fabric Structure: Satin
Fabric name: Sateen
Count: 64 × 104
Coloration: ________________
Weight: 6.65 oz./yd.^{2}
Uses: Bedding, upholstery, accessories

Characteristics

Sateen is the cotton variation of a satin. The luster is lower and richer due to the spun yarns. The fabric is heavier and denser.

Sateen is traditionally made with floats in the weft direction, although there are many exceptions to this rule.

Satin Weaves

Swatch 67

Swatch 68

Fiber content: Cotton
Yarn construction: Spun
Fabric structure: Slack-tension weave
Fabric name: Seersucker
Count: 64 × 50
Coloration: Yarn dyed
Weight: 6.95 oz./yd.^{2}
Uses: Light upholstery, windows, accessories

Characteristics

Seersucker is a slack tension weave. Note that the structure is a simple plain weave, yet groups of warp yarns are puckered; slack tension means that some of the yarns are held at a slack, or looser, tension compared with the rest of the yarns. This causes the vertical puckering that is integral to seersucker. This lightweight fabric is easy to care for, as it is a cotton fabric that requires no ironing. Seersucker can be yarn dyed, piece dyed, or printed.

Swatch 69

Fiber content: 76% Rayon/24% lyocell
Yarn construction: Filament
Fabric structure: Dobby
Fabric name: Dobby squares
Count: 104 × 80
Coloration: Piece dyed
Weight: 1.97 oz./yd.^{2}
Uses: Window treatments

Characteristics

Dobby structures are made up of small, hard-edged, geometric motifs created by two different weave structures combined in the fabric. This creates a foreground and background in the woven design.

Slack-tension Weave

Swatch 68

Dobby Weaves

Swatch 69

Dobby Weaves

Swatch 70

Swatch 70

Fiber content: Cotton
Yarn construction: Spun
Fabric structure: Dobby
Fabric name: Honeycomb
Count: 104 × 48
Coloration: Print
Weight: 7.37 oz./yd.^{2}
Uses: Upholstery, tableware, kitchen towels

Characteristics

Dobby structures, such as honeycombs, waffles, and piqués, are well known for their superior resilience. The small design adds interest to the surface of the fabric.

Dobby Weaves

Swatch 71

Swatch 71

Fiber content: Cotton
Yarn construction: Spun
Fabric structure: Dobby
Fabric name: Bird's-eye piqué
Count: 132 × 54
Coloration: Print
Weight: 5.94 oz./yd.^{2}
Uses: Upholstery, tableware, kitchen towels

Characteristics

Bird's-eye piqué is a specific dobby design. It can be piece dyed, yarn dyed, or printed. The structure has excellent resilience and dimensional stability.

Swatch 72

Fiber content: Cotton
Yarn construction: Spun
Fabric structure: Dobby
Fabric name: Waffle cloth
Count: 132 × 80
Coloration: Print
Weight: 3.52 oz./$yd.^2$
Uses: Towels, accessories, light upholstery
Ⓢ Ⓡ Ⓑ

Characteristics

Waffle cloth has been a favorite for kitchen towels due to its superior absorbency. This waffle is smaller, delicate, and more decorative and could be used for accessories and even lightweight, low-use upholstery.

Swatch 73

Fiber content: Cotton
Yarn construction: Two-ply spun
Fabric structure: Dobby
Fabric name: Goose-eye twill
Count: 38 × 20
Coloration: Yarn dyed
Weight: 8.87 oz./$yd.^2$
Uses: Upholstery, pillows
Ⓢ Ⓡ Ⓑ

Characteristics

Goose-eye twill is an old standard fabric used for upholstery and decorative accents due to the graphic nature of the design.

Dobby Weaves

Swatch 72

Dobby Weaves

Swatch 73

Dobby Weave

Swatch 74

Swatch 74

Fiber content: Polyester
Yarn construction: Filament
Fabric structure: Dobby
Fabric name: Satin stripe
Count: 162 × 68
Coloration: Yarn dyed
Weight: 4.17 oz./yd.^{2}
Uses: Light upholstery, accessories
Ⓡ

Characteristics

Striped satin is a dobby because the design is made of two different weaves. The stripes are a raised satin, while the ground is recessed in a plain weave.

Dobby Weave

Swatch 75

Swatch 75

Fiber content: Rayon
Yarn construction: ____________
Fabric structure: Dobby
Fabric name: Momie or crepe weave
Count: 68 × 50
Coloration: Piece dyed
Weight: 9.44 oz./yd.^{2}
Uses: Light upholstery (needs backing), accessories
Ⓢ Ⓡ Ⓑ

Characteristics

Crepe is a good choice for upholstery, although it should be heavier weight or backed. The irregular, textured pattern absorbs light and hides traffic patterns. It also has superior resilience, drape, and elastic recovery.

Swatch 76

Fiber content: Cotton
Yarn construction: Spun
Fabric structure: Extra-yarn weave
Fabric name: Extra-yarn weave
Count: 68 × 42
Coloration: Yarn dyed
Weight: 4.82 oz./yd.²
Uses: Kitchen and tableware

Characteristics

Extra-yarn weaves allow color to be introduced to the fabric in areas not possible in other weave structures. An extra set of yarns floats behind the fabric and comes forward to weave the motif when required. This results in more complex color systems than is available with only two sets of yarns.

Swatch 77

Fiber content: Cotton
Yarn construction: Spun
Fabric structure: Extra-yarn weave
Fabric name: Clip spot
Count: 82 × 64
Coloration: Piece dyed
Weight: 2.36 oz./yd.²
Uses: Curtains

Characteristics

A clip spot is a type of extra-yarn weave, where the floats are clipped off after weaving. The pattern, or motif, looks like individual embroideries. More complex colorways are possible with these weave types.

Extra-yarn Weave

Swatch 76

Extra-yarn Weave

Swatch 77

Swatch 78

Swatch 78

Fiber content: Cotton
Yarn construction: Spun
Fabric structure: Extra-yarn weave
Fabric name: Dotted swiss
Count: 100 × 68
Coloration: Piece bleached
Weight: 2.4 oz./yd.^{2}
Uses: Curtains

Characteristics

Dotted swiss is a variation of a clip spot; the pattern is dots or spots. The floats are clipped off and the fabric is used with the clipped spots on the front. The fabric is also called dot/spot or clipped dot. Flocked dotted swiss (Swatch 132) is a common knockoff.

Swatch 79

Fiber content: Rayon/polyester
Yarn construction: ______________
Fabric structure: Double-weave jacquard
Fabric name: Double-weave chenille tapestry
Count: 120 × 64 (each face)
Coloration: Yarn dyed
Weight: 14.46 oz./yd.^{2}
Uses: Upholstery

Characteristics

Tapestry is the simplest weave construction of the jacquards, usually made of a single structure with many colors. This variation is also a double weave, made with four sets of yarns. The yarns include chenille for a warm, soft, and fuzzy hand. Historic tapestries are made with a discontinuous weft and are extremely labor intensive. This jacquard variation is quite popular today.

Swatch 80

Fiber content: Cotton
Yarn construction: Spun
Fabric structure: Jacquard
Fabric name: Damask
Count: 172 × 94
Coloration: Piece bleached
Weight: 4.22 oz./yd.^{2}
Uses: Table linens, bedding, windows
Ⓢ Ⓡ Ⓑ

Characteristics

Damask, named for its origins in Damascus, was made famous as it was carried to the rest of the world along the Silk Road. The lightest of the jacquards, this fabric is easily recognizable, as front and back are mirror image reversible. This is accomplished by using satin floats at right angles to one another to create the pattern. Damask can reverse with either a shift in luster from face to back (as in this one) or a reversal of colors when made of two colors. Cotton damasks have long been called tablecloth damask, although this fine example from France would be as appropriate for window treatments and bedding.

Jacquard Weaves

Swatch 79

Jacquard Weaves

Swatch 80

Jacquard Weaves

Swatch 81

Swatch 81

Fiber content: Cotton/polyester

Yarn construction: Warp—Filament
Weft—Spun

Fabric structure: Jacquard

Fabric name: Damask

Count: 104 × 34

Coloration: Print

Weight: 6.14 oz./yd.^{2}

Uses: Upholstery, bedding, tableware

Characteristics

The structure of the damask is still reversible, but as the fabric is printed, it does not appear so. This damask is heavier than the previous example but still quite light for upholstery on its own. Backing would be required for durability.

Jacquard Weaves

Swatch 82

Swatch 82

Fiber content: Cotton

Yarn construction: Two-ply spun

Fabric structure: Jacquard

Fabric name: Flamestitch damask

Count: 168 × 70

Coloration: Yarn dyed

Weight: 10.0 oz./yd.^{2}

Uses: Upholstery

Characteristics

This heavier damask is a good weight for upholstery. Notice that the colors shift from front to back—they are opposite. The pattern or design is a flamestitch and could be rendered in a tapestry, brocade, damask, or even a print.

Swatch 83

Fiber content: Nylon/polyester
Yarn construction: Filament
Fabric structure: Jacquard
Fabric name: Brocade
Count: 132 × 68
Coloration: Yarn dyed
Weight: 9.08 oz./yd.^{2}
Uses: Windows, upholstery, bedding

Characteristics

Brocades were once made with actual silver and gold. Today, they are often made with metallics along with filament yarns on a jacquard loom. This fabric is not reversible and generally has a smaller repeat than tapestry (Swatch 79). Brocade uses a third set of yarns in the weft direction to create the pattern. Brocade is characterized by figurative designs, and is the most complicated of the jacquards in terms of structure and color. This fabric can be constructed with multiple weave structures.

Swatch 84

Fiber content: Polyester
Yarn construction: Spun & chenille yarns
Fabric structure: Jacquard
Fabric name: Brocatelle
Count: 172 × 100
Coloration: Yarn dyed
Weight: 17.79
Uses: Upholstery, windows, bedding

Characteristics

Brocatelle is essentially the same as brocade but heavier and made of spun yarns. For this reason, the fabric is usually more three-dimensional than brocade.

Jacquard Weaves

Swatch 83

Jacquard Weaves

Swatch 84

Swatch 85

Swatch 85

Fiber content: Polyester
Yarn construction: Filament
Fabric structure: Jacquard
Fabric name: Liseré
Count: 108 × 56
Coloration: Yarn dyed
Weight: 5.94 oz./yd.^{2}
Uses: Upholstery, bedding

Characteristics

Liseré is a French brocade with a very specific pattern. The design has stripes alternating with vertical rows of flowers (flower stripes). Traditionally a silk fabric made with brocaded flowers, the flowers are made with the extra brocading yarns, floating on the back when not in use.

Swatch 86

Fiber content: Polyester/rayon
Yarn construction: Warp—Filament
Weft—Filament, spun & two-ply spun
Fabric structure: Double weave
Fabric name: Double-weave jacquard
Count: 128 × 100 (each face)
Coloration: Yarn dyed
Weight: 10.66 oz./yd.^{2}
Uses: Upholstery, bedding, windows

Characteristics

Double weaves are essentially two fabrics woven together at the same time (as opposed to being sewn together at a later date). Double weaves can be made with four sets of yarns, as this one is, or five sets of yarns. In this fabric, the warp and weft exchange places, top to bottom, to create pockets. Where a fifth set of yarns are used, two separate fabrics are woven at the same time and the fifth yarn weaves the two together, without exchanging places.

Swatch 87

Fiber content: Cotton
Yarn construction: Single & two-ply spun
Fabric structure: Double weave
Fabric name: Double-weave pocket cloth
Count: 116 × 116 (each face)
Coloration: Piece bleached
Weight: 6.74 oz./yd.^{2}
Uses: Bedding
Ⓢ Ⓡ Ⓑ

Characteristics

As the two layers of the double weave interchange, they create equal pockets in the fabric structure. The face of the fabric actually looks like a quilt with its diamond design but is actually a good example of a double-weave structure.

Double Weaves

Swatch 86

Double Weaves

Swatch 87

Swatch 88

Swatch 88

Fiber content: Cotton
Yarn construction: Two-ply spun
Fabric structure: Double weave
Fabric name: Cloqué
Count: 104 × 64 (front)
Coloration: Piece dyed
Weight: 11.96 oz./yd.^{2}
Uses: Upholstery, mattress pads
Ⓢ Ⓡ Ⓑ

Characteristics

Cloqué is a double weave that is more dimensional than the previous examples. The fabric generally has some kind of padding or stuffing between the layers, (stuffer yarns) causing a puckered or embossed appearance.

Double Weaves

Swatch 89

Swatch 89

Fiber content: Polyester/rayon
Yarn construction: Filament
Fabric structure: Double weave
Fabric name: Matelassé
Count: 156 × 80 (face)
Coloration: Yarn dyed
Weight: 5.72 oz./yd.^{2}
Uses: Bedding, windows

Characteristics

Matelassé is a French word that references its *blistered* surface. This fabric is a double weave that is often made with different fibers or different yarn types to encourage the blistering between layers. Essentially, one layer shrinks, the other puckers, and the fabric becomes three-dimensional. Woven with four sets of yarns.

Swatch 90

Fiber content: Polyester
Yarn construction: Filament with some slubs
Fabric structure: Leno
Fabric name: Marquisette
Count: 60 × 12
Coloration: Piece dyed
Weight: 2.42 oz./yd.^{2}
Uses: Window sheers

Characteristics

A leno weave wraps pairs of warp yarns around the weft yarns in a figure-8 pattern. This allows the structure to remain light and ephemeral, without risk of the yarns shifting or the ultimate collapse of the weave structure. This is the best choice for an open construction woven lace fabric.

Leno Weave

Swatch 90

Swatch 91

Fiber content: Cotton
Yarn construction: Spun
Fabric structure: Pile weave
Fabric name: Terry velour
Count: 72 × 46
Coloration: ________________
Weight: 8.93 oz./yd.^{2}
Uses: Towels
Finishes: Shearing, napping
Ⓢ Ⓡ Ⓑ

Characteristics

Terry cloth is a pile structure, which means that it uses an extra set of yarns to create a three-dimensional fabric surface. Terry velour means that the loops are cut on one side for a smoother, more luxurious face and left uncut on the back. The uncut loops are much more absorbent, able to absorb many times its own weight in moisture. Thread count and weight play an important role in the quality of the towel.

Swatch 92

Fiber content: Cotton
Yarn construction: Spun
Fabric structure: Pile weave
Fabric name: Widewale corduroy
Count: 70 × 24
Coloration: Piece dyed
Weight: 15.24 oz./yd.^{2}
Uses: Upholstery
Finishes: Shearing, napping
Ⓢ Ⓡ Ⓑ

Characteristics

Corduroy is a cut pile made of an extra set of filling or weft yarns. The pile structure is arranged in vertical rows. Corduroys are identified by the number of rows or wales per inch, ranging from a pinwale corduroy (16–18 wales per inch), midwale (10–14 wales per inch) and a wide wale, such as this one.

Pile Weaves

Swatch 91

Pile Weaves

Swatch 92

Swatch 93

Swatch 93

Fiber content: Cotton
Yarn construction: Spun
Fabric structure: Pile weave
Fabric name: Velveteen
Count: 74 × 48
Coloration: Piece dyed
Weight: 9.33 oz./yd.^{2}
Uses: Upholstery
Finishes: Shearing, napping
Ⓢ Ⓡ Ⓑ

Characteristics

Velveteen also uses a third set of yarns to make loops on the surface of the fabric. The loops are sheared and napped or brushed to create a soft, plush surface. Velveteen differs from velvet in that it is made of spun yarns, usually cotton. This provides velveteen with more body and less drape and luster.

Swatch 94

Swatch 94

Fiber content: Nylon
Yarn construction: Filament
Fabric structure: Pile weave
Fabric name: Velvet
Count: 88 × 80
Coloration: Piece dyed
Weight: 4.72 oz./yd.^{2}
Uses: Light upholstery, windows, bedding
Finishes: Shearing, napping
Ⓡ

Characteristics

Velvet is smoother, softer, and more lustrous, due to the fact that it is made of filament yarns. This one is not heavy enough for upholstery, but it can be used as a window treatment. Velvet is generally a shorter pile than velveteen.

Swatch 95

Fiber content: Polyester
Yarn construction: Filament
Fabric structure: Pile weave
Fabric name: Crushed velvet
Count: 84 × 70
Coloration: Piece dyed
Weight: 10.4 oz./yd.^{2}
Uses: Upholstery, windows, accessories, bedding
Finishes: Shearing, napping

Characteristics

Panné or crushed velvet adds an extra step in the finishing process, literally crushing the pile (in crushed velvet) or laying it flat in one direction, as in Panné velvet.

Swatch 96

Fiber content: Rayon/polyester
Yarn construction: Warp—Two-ply spun
Weft—Filament
Fabric structure: Pile weave
Fabric name: Voided velvet
Count: 60 × 50
Coloration: Yarn dyed
Weight: 13.11 oz./yd.^{2}
Uses: Upholstery, windows
Finishes: Shearing, napping

Characteristics

Voided velvet is a patterned velvet that is yarn dyed. Different from embossed designs in velvet, this is a complicated, intricate, woven velvet design.

Pile Weaves

Swatch 95

Pile Weaves

Swatch 96

Pile Weave

Swatch 97

Swatch 97

Fiber content: ____________________
Yarn construction: Two-ply spun
Fabric Structure: Pile weave
Fabric name: Grospoint
Count: 90 × 52
Coloration: Piece dyed
Weight: 31.02 oz./yd.^{2}
Uses: Heavy upholstery
S R B

Characteristics

Grospoint is a heavy duty pile weave made with uncut loops, similar to frieze. The warp and weft are completely covered.

Swatch 98

Fiber content: Cotton
Yarn construction: Spun
Fabric structure: Weft knit
Fabric name: Jersey
Count: 37 × 54
Coloration: Stock dyed
Weight: 4.17 oz./yd.^{2}
Uses: Bedding
Ⓢ Ⓡ Ⓑ

Characteristics

Jersey is the simplest of weft knits and stretches easily in both directions. This makes stretching the fabric over a mattress easy to accomplish. For this reason, it is quite desirable for bedding. Additionally, wrinkles are stretched out. Finally, the soft and cozy hand of a cotton jersey can also be a good selling point.

Swatch 99

Fiber content: ____________________
Yarn construction: Filament
Fabric structure: Double-weft knit
Fabric name: Double-knit mattress pad
Count: 26 × 11 (face)
Coloration: Piece bleached
Weight: 6.35 oz./yd.^{2}
Uses: Mattress pad
Ⓡ

Characteristics

A double knit is very stable and keeps its shape better than a jersey. It is also soft and yielding, once again, making it easier to stretch over a mattress. This one has a central padding or stuffing, similar to a woven cloqué or matelassé to assist in keeping its three-dimensional quality.

Weft Knits

Swatch 98

Double-weft Knits

Swatch 99

Pile Knits

Swatch 100

Swatch 100

Fiber content: Acrylic
Yarn construction: Spun & sliver
Fabric structure: High pile knit
Fabric name: Sliver knit (faux fur)
Count: 18 × 26
Coloration: Fiber dyed
Weight: 17.44 oz./yd.^{2}
Uses: Pillows, bedding, accessories

Characteristics

Sliver knit does not catch an extra set of yarns into each stitch, making loops on the surface that are later sheared (as the other pile structures do). Instead, sliver is caught into each stitch. Sliver is cleaned, carded fiber. Animal designs and faux fur are used for pillows and accessories in the home market today. This design is called African leopard.

Pile Knits

Swatch 101

Swatch 101

Fiber content: Polyester
Yarn construction: Filament
Fabric structure: Pile knit
Fabric name: Fleece
Count: 34 × 56
Coloration: ____________
Weight: 7.47 oz./yd.^{2}
Uses: Bedding
Finishes: Napping

Characteristics

Fleeces are soft and warm, very appropriate for bedding. Fleece is napped on both sides, and sweatshirt fleece is napped on one side only. Usually made of polyester, this fabric is also called double-sided fleece.

Swatch 102

Fiber content: Polyester
Yarn construction: Filament
Fabric structure: Warp knit bonded to a weft knit
Fabric name: Bonded and embossed pile knit
Count: 16 × 15 (face)
Coloration: Yarn dyed
Weight: 15.2 oz./yd.^{2}
Uses: Upholstery
Finishes: Napping, embossing

Characteristics

The texture in this pile knit is caused by both embossing and bonding two knits together. This provides great compressional resilience for an upholstery fabric. The face is a warp knit and the back is an embossed single faced fleece.

Pile Knit

Swatch 102

Swatch 103

Fiber content: Nylon
Yarn construction: Filament
Fabric structure: Tricot
Fabric name: Tricot
Count: 76 × 108
Coloration: Piece dyed
Weight: 6.2 oz./yd.^{2}
Uses: Event chair covers

Characteristics

Tricot is the simplest of warp knits. The greatest advantage is that warp knits do not run or ravel, although the filament yarns do have a propensity to snag. This fabric can be used for chair covers for special events due to their ease in removal and replacement.

Swatch 104

Fiber content: Polyester
Yarn construction: Filament
Fabric structure: Tricot
Fabric name: Brushed tricot
Count: 42 × 52
Coloration: Piece dyed
Weight: 5.68 oz./yd.^{2}
Uses: Light upholstery
Finishes: Napping

Characteristics

Brushed tricot is a tricot with a napped finish. The wire bristles of the napping process break just enough fibers to create a soft and fuzzy surface. This fabric can be found in automotive headliners, airplane blankets, and small upholstered pieces.

Warp Knits

Swatch 103

Warp Knits

Swatch 104

Warp Knits

Swatch 105

Swatch 105

Fiber content: Polyester
Yarn construction: Filament
Fabric structure: Tricot
Fabric name: Sueded tricot
Count: 32 × 40
Coloration: Piece dyed
Weight: 4.8 oz./yd.²
Uses: Light upholstery
Finishes: Emerizing

Characteristics

Sueded tricot is one of many imitation suedes on the market today, often called vegan suede. It is a very stable knit made of polyester and so is quite durable and stain resistant.

Warp Knits

Swatch 106

Swatch 106

Fiber content: Polyester
Yarn construction: Filament
Fabric structure: Tricot
Fabric name: Minky (rose tricot)
Count: 36 × 24
Coloration: Piece dyed
Weight: 15.09 oz./yd.²
Uses: Blankets, pillows, accessories

Characteristics

The "rose tricot" is essentially a pile or sliver tricot, in that it incorporates an extra set of fibers to create a third dimension. The industry calls these minkies due to their very soft, mink-like hand. Minkies have been replacing traditional flannels and flannelettes in the apparel market. Made on a tricot body, this is a more contemporary variation of tufting.

Swatch 107

Fiber content: Nylon
Yarn construction: Filament
Fabric structure: Raschel
Fabric name: Tulle
Count: 8 × 7
Coloration: Piece dyed
Weight: .83 oz./yd.^{2}
Uses: Sheers, bed skirts & canopies, decorative elements
Finishes: Stiffening

Characteristics

Tulle has a netlike structure with diamond or hexagonal openings and a stiffened finish, like organdy (Swatch 40). It is often used in layers to create volume and ruffles.

Swatch 108

Fiber content: Nylon
Yarn construction: Filament
Fabric structure: Raschel
Fabric name: Raschel lace
Count: 8 × 10
Coloration: ______________
Weight: 1.09 oz./yd.^{2}
Uses: Windows, tableware

Characteristics

Raschel is a lacelike structure made on a hex net or tulle ground. This is the most common lace type in the marketplace and has many variations in terms of weight and construction.

Warp Knits

Swatch 107

Warp Knits

Swatch 108

Warp Knits

Swatch 109

Swatch 109

Fiber content: Polyester
Yarn construction: Textured filament
Fabric structure: Raschel
Fabric name: Raschel crochet
Count: 22 × 16
Coloration: Piece dyed
Weight: 6.92 oz./yd.^{2}
Uses: Windows, tableware

Characteristics

Raschel crochet can be made with cotton or a textured polyester. The fabric is made on a Raschel machine but is meant to imitate a handmade cotton crochet, which is an actual lace.

Warp Knits

Swatch 110

Swatch 110

Fiber content: Nylon/spandex
Yarn construction: Filament
Fabric structure: Raschel
Fabric name: Point d'esprit
Count: 18 × 12
Coloration: Piece bleached
Weight: 1.64 oz./yd.^{2}
Uses: Window sheers

Characteristics

Point d'esprit is an old millinery fabric that lends itself quite well to window sheers. The fabric is the knit industry's answer to a woven dotted swiss and is sometimes called spot knit for this reason.

Swatch 111

Fiber content: 90% Polyester/10% polyurethane
Yarn construction: No yarns
Fabric structure: Nonwoven
Fabric name: Imitation or Vegan suede
Count: No count
Coloration: Piece dyed
Weight: 11.2 oz./yd.^{2}
Uses: Upholstery

Characteristics

This is a performance fabric that offers a soft hand, moisture absorption, quick drying, and excellent durability. Imitation suede is a nonwoven construction that successfully imitates actual suede, without the need for an animal. A common trade name for this type of structure is Ultrasuede®, and sometimes it is quite difficult to discern the imitation from actual suede. A burn test is required. The fabric is a good substitute because it is cruelty-free, washable, stain resistant, and infinitely easier to care for. Notice the lack of grain.

Swatch 112

Fiber content: Recycled Eco-fi polyester
Yarn construction: No yarns
Fabric structure: Nonwoven
Fabric name: Needlepunched felt
Count: No count
Coloration: Piece dyed
Weight: 4.75 oz./yd.^{2}
Uses: Upholstery, backings

Characteristics

Needlepunched or imitation felt is made by the entanglement process. Actual felt is the matting of hair, but it requires at least 50% animal hair (wool), not present in this swatch. Felt-type fabrics are used as carpet underlay, light upholstery, bedding, and accessories. As there is no grain in either of these fabrics, they can be molded over a form conforming equally in both directions.

Nonwovens

Swatch 111

Nonwovens

Swatch 112

Film

Swatch 113

Film

Swatch 114

Swatch 113

Fiber content: Polyurethane (PU)/vinyl
Yarn construction: No yarns
Fabric structure: Bonded
Fabric name: Pleather
Count: No count
Coloration: Print
Weight: 8.84 oz./yd.^{2}
Uses: Upholstery
Finishes: Flame retardant, embossing

Characteristics

Pleather is sometimes called vegan leather and is a good alternative to leather as it is easier to care for and is quite durable due to its fiber content. There are also many more colors and finishes possible. Often, a pleather has a printed and embossed film on the face, with a knit or nonwoven substrate bonded to the back.

Swatch 114

Fiber content: Polyvinyl chloride (PVC)/polyester/cotton
Yarn construction: Filament
Fabric structure: Film
Fabric name: Oilcloth
Count: 20 × 12
Coloration: ______________
Weight: 8.45 oz./yd.^{2}
Uses: Outdoor tablecloth, furniture

Characteristics

Oilcloth is an old term for a canvas fabric that is coated with linseed oil. Today's oilcloth is often made of polyvinyl chloride (PVC), which makes the fabric water and stain resistant and appropriate for outdoor use in table covers and light furniture. Sometimes bonded to a nonwoven, this one has a woven grid embedded in the structure for strength and durability.

Swatch 115

Fiber content: Nylon face, Polyester batting and back
Yarn construction: Filament
Fabric structure: Fabric combinations
Fabric name: Quilt
Count: 120 × 96
Coloration: Piece dyed
Weight: 8.76 oz./yd.^{2}
Uses: Bedding, upholstery, pillows

Characteristics

A quilted fabric is three fabric layers held together by a decorative stitch. Traditionally, the face is woven, the middle layer is a nonwoven batting, and the back is a simple muslin. The back of this one is a knit. The three-dimensional structure offers thermal retention and padding to a surface.

Swatch 116

Fiber content: Cotton
Yarn construction: Spun
Fabric structure: Plain weave
Fabric name: Embroidered eyelet
Count: 90 × 82
Coloration: Piece bleached
Weight: 3.75 oz./yd.^{2}
Uses: Windows, tableware, bedding
Ⓢ Ⓡ Ⓑ

Characteristics

Embroidery is a surface embellishment stitched onto the surface of an existing fabric. Holes are often placed in the fabric, and the fabric is then called an embroidered eyelet. There are many embroideries that are not eyelets. Embroideries can be added to the surface of any fabric.

Fabric Combinations

Swatch 115

Embroideries

Swatch 116

Tufted

Swatch 117

Swatch 117

Fiber content: Cotton
Yarn construction: Spun
Fabric structure: Tufting
Fabric name: Tufted chenille
Count: 60 × 68
Coloration: Piece dyed
Weight: 10.57 oz./yd.²
Uses: Bedding
Ⓢ Ⓡ Ⓑ

Characteristics

Tufted chenille (sometimes called candlewick) is a plain-weave cotton fabric. Yarns are inserted into the woven fabric, much like carpeting. Patterns are created with the tufting.

Lace

Swatch 118

Swatch 118

Fiber content: Cotton
Yarn construction: Spun
Fabric structure: Lace
Fabric name: Venise lace (guipure)
Count: No count
Coloration: Piece bleached
Weight: 6.08 oz./yd.²
Uses: Tableware, windows, bedding
Ⓢ Ⓡ Ⓑ

Characteristics

Real lace is neither woven nor knit; it is made by many other processes, such as knotting, looping, or twisting. Guipure is an embroidered type of lace, created on a fabric that dissolves in water after the embroidery is complete. Effectively, this is embroidery “in the air” since the foundation fabric is gone.

Swatch 119

Fiber content: Cow
Yarn construction: No yarn
Fabric structure: Skin
Fabric name: Leather
Count: No count
Coloration: Piece dyed
Weight: 27.55 oz./yd.^{2}
Uses: Upholstery
Finishes: Tanning

Characteristics

Leather is the outside layer of the skin of an animal, usually, but not always, cattle. The inside of the skin is called suede. Leather can be quite luxurious and desirable but is significantly more expensive and can be more difficult to care for. Leather should be oiled and protected from the sun.

Skin

Swatch 119

Swatch 120

Fiber content: Polyester
Yarn construction: Filament with white slubs
Fabric structure: Twill
Fabric name: Herringbone tweed
Count: 84 × 68
Coloration: Appears stock dyed
Weight: 7.86 oz./yd.^{2}
Uses: Upholstery

Characteristics

This herringbone tweed appears stock dyed, (which means fiber dyed) but is not made of staple fibers. The fabric does have the distinctive heathered or tweedy appearance. Remove a yarn to notice that it has color variations along its length, along with distinctive slubs.

Swatch 121

Fiber content: Polyester
Yarn construction: Warp—Filament
Weft—Spun
Fabric structure: Jacquard
Fabric name: Damask
Count: 104 × 36
Coloration: Yarn dyed
Weight: 7.77 oz./yd.^{2}
Uses: Upholstery, drapery, bedding

Characteristics

Yarn dyed means that the yarns were dyed, and colored yarns were brought together to weave this fabric. Remove a few yarns to discover that each yarn is a different color. In this fabric, the different colored yarns are in different directions on the fabric. Jacquard patterns, along with checks and plaids, are usually yarn dyed.

Stage of Dyeing

Swatch 120

Stage of Dyeing

Swatch 121

Stage of Dyeing

Swatch 122

Swatch 122

Fiber content: Cotton
Yarn construction: Spun
Fabric structure: ______________
Fabric name: ______________
Count: 116 × 100
Coloration: Piece dyed
Weight: 8.48 oz./yd.^{2}
Uses: Upholstery, slipcovers, bedding
Ⓢ Ⓡ Ⓑ

Characteristics

Piece dyeing results in a solid color, since the whole fabric has been dyed at once. Although other stages of dyeing (yarn dyeing, for example) could result in a solid color, this effect is usually achieved by piece dyeing, as it is less expensive than earlier stages in the production cycle.

Classic Print Designs

Swatch 123

Swatch 123

Fiber content: Pima cotton
Yarn construction: Spun
Fabric structure: Plain weave
Fabric name: Calico/ditsy print
(Also known as a liberty print)
Count: 164 × 104
Coloration: Print
Weight: 2.4 oz./yd.^{2}
Uses: Windows, bedding, accessories
Ⓢ Ⓡ Ⓑ

Characteristics

Calico is one of the earliest printed fabrics and comes from the city of Calcutta in India. It is a lightweight, plain-weave, cotton fabric with a small floral print. Today, these fabrics are often called ditsy (a small print). The design is printed on a lightweight Pima cotton, plain weave.

Swatch 124

Fiber content: Cotton
Yarn construction: Spun
Fabric structure: Half basket
Fabric name: Paisley print (on canvas)
Count: 64 × 28
Coloration: Print
Weight: 7.17 oz./yd.^{2}
Uses: Light upholstery, accessories
S R B

Characteristics

Paisley was originally called Kashmir, for the city of the same name. The earliest paisleys were tapestries and were (and are) highly collectable. Today, paisleys are most often printed. The pattern has a distinctive shape based on the original boteh, or drooping flower.

Classic Print Designs

Swatch 124

Swatch 125

Fiber content: Cotton
Yarn construction: Spun
Fabric structure: Plain weave
Fabric name: Toile de Jouy
Count: 92 × 60
Coloration: Print
Weight: 4.78 oz./yd.^{2}
Uses: Bedding, wallcoverings, accessories, pillows
S R B

Characteristics

Toile de Jouy was an early French printed fabric. The prints were genre or pastoral scenes, often commemorating current events in French society. The print is traditionally a large monochromatic repeat, printed with fine detail on plain or basket-weave cotton or linen. This contemporary variation has more than one color.

Classic Print Designs

Swatch 125

Non-Classic Designs

Swatch 126

Swatch 126

Fiber content: Cotton/flax
Yarn construction: Spun
Fabric structure: Plain weave
Fabric name: "Linen look"
Count: 60×46
Coloration: Direct print
Weight: 4.59 oz./yd.^{2}
Uses: Light upholstery, bedding

Characteristics

In a direct print, the foreground is printed on a white (or undyed) background. This is the simplest and least expensive print, using one less screen, as the background is not printed.

Non-Classic Designs

Swatch 127

Swatch 127

Fiber content: Flax/rayon
Yarn construction: Spun
Fabric structure: ____________
Fabric name: ____________
Count: ____________
Coloration: ____________
Weight: 5.67 oz./yd.^{2}
Uses: Light upholstery, bedding, windows

Characteristics

Blotch print uses one more screen or roller than a direct print, as the background is printed.

Swatch 128

Fiber content: Cotton/spandex
Yarn construction: Spun
Fabric structure: Pile weave
Fabric name: Velveteen
Count: 112 × 64
Coloration: Overprint
Weight: 8.26 oz./yd.^{2}
Uses: Upholstery, bedding
Finishes: Shearing, napping

Characteristics

An overprint also has a background color but has been dyed prior to printing. The fabric is then printed *over* the dyed fabric. Notice that the background color is solid on the front and back of the fabric.

Swatch 129

Fiber content: Rayon
Yarn construction: Filament
Fabric structure: Rib weave
Fabric name: Crepe faille
Count: 104 × 66
Coloration: Discharge print
Weight: 3.18 oz./yd.^{2}
Uses: Windows
Ⓢ Ⓡ Ⓑ

Characteristics

Discharge is a two-step process. First, the fabric is dyed, usually a dark color. Second, the fabric is printed with a paste that discharges or removes color. The next roller lays color back in. It is easy to recognize a discharge print because the print is equally visible on both sides.

Non-Classic Designs

Swatch 128

Better Quality Prints

Swatch 129

Better Quality Prints

Swatch 130

Better Quality Prints

Swatch 131

Swatch 130

Fiber content: 97% Cotton/3% spandex
Yarn construction: Spun
Fabric structure: Twill
Fabric name: Flocked Denim
Count: 88 × 68
Coloration: Yarn dyed, flock printed
Weight: 8.99 oz./yd.^{2}
Uses: Upholstery, accessories

Characteristics

A flock print is initially printed with glue, and then tiny, short flocks are electrostatically adhered to the surface. Flock prints have fuzzy, dimensional designs.

Swatch 131

Fiber content: Nylon
Yarn construction: Filament
Fabric structure: Plain weave
Fabric name: Flocked dotted swiss
Count: 90 × 90
Coloration: Flock print
Weight: .77 oz./yd.^{2}
Uses: Curtains

Characteristics

Flocked dotted swiss is a flock print intended to imitate the clip spot, dotted swiss. This one is on an organza body.

Swatch 132

Fiber content: Silk/rayon
Yarn construction: Filament
Fabric structure: Pile weave
Fabric name: Burnout velvet
Count: 108 × 112
Coloration: Burnout
Weight: 2.99 oz./yd.^{2}
Uses: Curtains
Finishes: Shearing, napping

Characteristics

Burnout is made with a fabric made of two different fibers, one that is a cellulose and one that is not. This velvet is made of silk (protein) and rayon (cellulosic). The fabric is printed with a chemical that destroys the cellulosic pile, leaving the silk ground weave intact. In this way, the rayon fibers are "burned out" of the fabric, creating a design.

Swatch 133

Fiber content: Polyester/rayon
Yarn construction: Warp—Filament
Weft—Spun & Filament
Fabric structure: Plain weave
Fabric name: Burnout
Count: 34 × 96
Coloration: Burnout & bleached
Weight: 1.99 oz./yd.^{2}
Uses: Curtains

Characteristics

This burnout also uses two fibers but is not a velvet. The weft has two sets of yarns; the larger rayon yarn is burned away, leaving a pattern of selective transparency.

Better Quality Prints

Swatch 132

Better Quality Prints

Swatch 133

Better Quality Prints

Swatch 134

Swatch 134

Fiber content: Polyester
Yarn construction: Filament
Fabric structure: ________________
Fabric name: ________________
Count: 86 × 44
Coloration: Duplex print
Weight: 2.02 oz./yd.^{2}
Uses: Curtains

Characteristics

Duplex print means that two sides of the fabric are printed with two different prints. Face and back are quite different for contrast. This would make a rather remarkable curtain, as one side could face the street and a different design would face the room.

Resist Prints

Swatch 135

Swatch 135

Fiber content: Cotton
Yarn construction: Spun
Fabric structure: Plain weave
Fabric name: Tie-dye
Count: 100 × 86
Coloration: Tie-dye
Weight: 1.94 oz./yd.^{2}
Uses: Curtains, accessories, bedding

Characteristics

Tie-dye is a generic term for a group of fabrics that are bound or tied in some way to resist the dye that is later applied. The fabric can be tied in concentric circles, sewn and pleated, or clamped; many variations are possible. Tie-dye has a more organic-appearing design than batik or ikat.

Swatch 136

Fiber content: Cotton
Yarn construction: Spun
Fabric structure: Plain weave
Fabric name: Batik
Count: 148 × 74
Coloration: Batik
Weight: 3.15 oz./yd.^{2}
Uses: Accessories, bedding

Characteristics

Batik is a resist print that uses wax to preserve predetermined areas of the fabric. A tjanting tool or stamp is used to apply the hot, liquid wax. The wax seals the fabric selectively and does not allow the dye to penetrate and color the fabric in protected areas. If the fabric has six colors, six different waxing and dye processes are required. Batik is easily distinguished by the cracked wax designs evident in the fabric design.

Resist Prints

Swatch 136

Swatch 137

Fiber content: Cotton
Yarn construction: Spun
Fabric structure: Plain weave
Fabric name: Ikat
Count: 42 × 30
Coloration: Ikat
Weight: 5.8 oz./yd.^{2}
Uses: Accessories, tableware, bedding

Characteristics

Ikat is accomplished by tying the individual threads for dyeing before weaving the fabric. This is the most arduous of the resist processes. Today, this fabric is often imported from Guatemala and Indonesia.

Resist Prints

Swatch 137

Swatch 138

Fiber content: Polyester
Yarn construction: Spun
Fabric structure: Twill
Fabric name: Flannel
Count: 24 × 30
Coloration: Yarn dyed
Weight: 10.15 oz./yd.^{2}
Uses: Blankets, bedding, light upholstery
Finishes: Napping
Ⓡ

Characteristics

Flannel is generally a twill fabric, used for flannel blankets or even light upholstery. Historically made of wool, flannels today can be made of polyester, cotton, rayon, or acrylic. The fabric is finished with the napping process, which is essentially brushing the fibers.

Swatch 139

Fiber content: Cotton
Yarn construction: Spun
Fabric structure: Plain weave
Fabric name: Flannelette
Count: 46 × 46
Coloration: Print
Weight: 3.93 oz./yd.^{2}
Uses: Sheets
Finishes: Napping
Ⓢ Ⓡ Ⓑ

Characteristics

Flannelette is a plain weave that is usually cotton and therefore lighter weight than flannel. This is a better choice for sheets.

Napping

Swatch 138

Napping

Swatch 139

Calendering

Swatch 140

Swatch 140

Fiber content: Cotton
Yarn construction: ______________
Fabric structure: Plain weave
Fabric name: Glazed Chintz
Count: 132 × 70
Coloration: Print
Weight: 3.29 oz./yd.²
Uses: Light upholstery, accessories
Finishes: Glazing

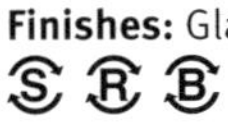

Characteristics

Glazed chintz is a plain-weave, printed, cotton fabric that has been polished and is slightly lustrous. The glazing finish is one of the calendering processes.

Calendering

Swatch 141

Swatch 141

Fiber content: Polyester
Yarn construction: Filament
Fabric structure: Rib weave
Fabric name: Moiré taffeta
Count: 96 × 44
Coloration: Piece dyed
Weight: 3.45 oz./yd.²
Uses: Windows, accessories
Finishes: Moiré

Characteristics

Moiré is a calendering process that impresses a wood grain design into a rib-weave structure. A moiré finish can also be applied to a bengaline.

Swatch 142

Fiber content: Polyester
Yarn construction: Filament
Fabric structure: Pile weave
Fabric name: Embossed crocodile velvet
Count: 42 × 48
Coloration: Piece dyed
Weight: 13.79 oz./yd.²
Uses: Windows, upholstery
Finishes: Shearing, napping, embossing
Ⓡ

Characteristics

Embossing a dimension into a fabric is another of the calendering processes. Any shape can be pressed into a three-dimensional fabric.

Swatch 143

Fiber content: Polyester
Yarn construction: ______________
Fabric structure: Dobby weave
Fabric name: Pleated georgette
Count: 168 × 116
Coloration: Piece dyed
Weight: 5.63 oz./yd.²
Uses: Windows
Finishes: Pleating
Ⓡ

Characteristics

If the fiber is thermoplastic (as polyester is), pleating can be accomplished with heat. If the fiber is not thermoplastic, chemicals must be used to ensure permanence.

Embossing

Swatch 142

Pleating

Swatch 143

Emerizing/Sueding

Swatch 144

Emerizing/Sueding

Swatch 145

Swatch 144

Fiber content: Polyester/rubber
Yarn construction: Filament
Fabric structure: Spacer fabric
Fabric name: Suede cloth
Count: 42 × 40
Coloration: Piece dyed
Weight: 11.62 oz./yd.^{2}
Uses: Upholstery
Finishes: Emerizing or sueding

Characteristics

This fabric is a spacer fabric, which means that there is a rubber foam construction sandwiched between two knits, followed by a sueded finish. This gives the fabric dimension.

Swatch 145

Fiber content: Polyester
Yarn construction: Filament
Fabric structure: Satin weave
Fabric name: Moleskin
Count: 120 × 78
Coloration: Piece dyed
Weight: 7.16 oz./yd.^{2}
Uses: Light upholstery
Finishes: Emerizing or sueding

Characteristics

Moleskin has a sueded finish on the back of a satin fabric. It can be folded back or reversed for contrast. Imitation suedes have many different permutations and possibilities today.